ESSAYS ON TEMPERATURE REGULATION

Essays on Temperature Regulation

Editors:
J. BLIGH
A.R.C. Institute of Animal Physiology
Babraham, Cambridge, Great Britain

R. E. MOORE
Department of Physiolology
Trinity College, Dublin, Ireland

1972

NORTH-HOLLAND PUBLISHING COMPANY — AMSTERDAM · LONDON
AMERICAN ELSEVIER PUBLISHING COMPANY, INC. — NEW YORK

ISBN North-Holland: 0 7204 4103 x
ISBN American Elsevier: 0 444 10346 5

Publishers:
NORTH-HOLLAND PUBLISHING COMPANY – AMSTERDAM
NORTH-HOLLAND PUBLISHING COMPANY, Ltd. – LONDON

Sole distributors for the Western Hemisphere:
AMERICAN ELSEVIER PUBLISHING COMPANY, INC.
52 VANDERBILT AVENUE, NEW YORK, N.Y. 10017

PRINTED IN THE NETHERLANDS

CONTENTS

PREFACE

This volume of 11 essays on various aspects of the control of body temperature is based on the invited lectures given on this theme at the International Symposium on Bioenergetics and Temperature Regulation held at Trinity College, Dublin in July 1971.

The titles of the essays were more or less imposed on the contributors, who were asked for their interpretations of recent studies in a number of broad fields of investigation. It was hoped that the writers would thus cover the subject in a manner halfway between a review of the literature and an account of their own researches.

The value of such assessments in any very active field of research depends entirely on their publication while still pertinent. To this end the Publisher recommended a rapid process of typescript reproduction, and the contributors were required to submit the typescript, including the figures, in a format that could be sent directly to the printer. Most of the contributors found difficulty in performing the tasks of both author and type-setter and did not quite succeed in presenting the required perfect copy. To have edited and corrected the typescripts would have involved returning them to the author and the original typist, and this we did not do. The minor variations in format, a few difficult passages which we might have persuaded the authors to change, and some unimportant typing errors seems preferable to the inevitable delay. We are not concerned about the overlaps of subject matter because each essay is a self-contained personal statement for which the author alone is responsible.

Thus the undersigned are nominal rather than active editors who played different roles in the preparation of the volume: one of us planned the series of lectures and persuaded the contributors to speak and then to write; the other organized the Dublin Symposium and thus created the occassion from which this volume has stemmed.

We acknowledge with gratitude the generous financial help of the Wellcome Trust and the Physiological Society (of Great Britain) which greatly eased the tasks of the programme and symposium organisers. Any praise for this volume must be shared by these benefactors, but any blame must be accepted by the editors.

We also wish to thank Dr K.G. Johnson for the production of the index, and the Publishers and the Contributors for their co-operation.

John Bligh

A.R.C. Institute of Animal Physiology, Babraham, Cambridge, England.

Roland E. Moore

Department of Physiology, Trinity College, Dublin, Ireland.

INTRODUCTION TO THE ESSAYS

Thermal physiology can be broadly divided into two compartments:
i) the physiology, biochemistry and physics of heat production and heat loss
which constitute the extra-central thermoregulatory effector functions, and ii)
the temperature sensors and central control processes by means of which the
balance between heat production and heat loss is held in that state of thermo-
dynamic equilibrium which can be expressed in terms of the stability of core
temperature. These essays chiefly relate to the latter compartment.

Much is known of what the brain achieves in relating disturbance or stimulus
to response, and in recent years something has been learned of the properties and
activities of individual neurons in the brain and the ways in which they interact
with other neurons, but still very little is known about how the central nervous
system achieves the innumerable, complex and largely reproducible relations between stimulus and response. The problems of investigating central nervous functions are further complicated by the fact that no functions of the body are really
wholly discrete. The brain is not a collection of separate "black-boxes" each
concerned with a particular function, but a single integrated piece of biological
machinery by which means all the functions of the body act to maintain the physiological integrity of the organism in the context of its total environment (i.e.
all the external influences which are acting on it).

The control of body temperature has attracted a great deal of interest, much
of which may have stemmed from our inevitable awareness of this particular home-
ostatic function since we devote much thought, money and energy (literally!) to
the creation of macro- and micro-climates in which we can maintain a state of
thermal comfort, and because of the clinical use of core temperature as an index
of health.

For the physiologist interested in control functions of the central nervous
system, thermoregulation has some advantages as a system for investigation; the
controlled quantity can be readily monitored, experimental disturbances in heat
production and in heat loss can be readily engineered, and the thermoregulatory
responses to these induced disturbances (sweating, panting, shivering, changes in
peripheral vasomotor tone) can be observed. It is recognised, of course, that the
control of body temperature is not really an isolated system, but under experimental conditions many of the other influences can be reduced or eliminated, and
thermoregulation can be studied almost as if it were.

So far the bulk of researches into the nature of the central nervous control
of body temperature relate to species which are considered to maintain fairly
stable core temperatures throughout their lives. As a general rule the experimental evidence indicates that the similarities in the processes of control out-
weigh the dissimilarities. The homeothermy of different species such as man, cat,
dog, rabbit, rat, monkey, sheep, goat and ox might thus be considered as variations
upon a common theme. A wider perusal of the literature reveals that the magnitude
of the species difference is really very large, and clearly a proper understanding
of the nature of the thermoregulation must take these differences between species
into account. Professor T.J. Dawson (New South Wales, Australia) was asked to
discuss some aspects of comparative thermoregulation in order to bring the subject
into its proper biological perspective. This is a large and involved field of
study and Professor Dawson was given little space to cover it. However, since
some earlier discussions on thermoregulation have implied that 'human thermoregulation' can be regarded in biological isolation, any light that can come from
the zoologists and comparative physiologists must help   the often narrow vision
of the applied biologists in the field of human medicine. Of course, not all mammals are permanently homeothermic. Some enter periods of daily or seasonal torpor,
others have wide cyclic nychthemeral variations in body temperature, and yet others
can maintain fairly stable body temperatures only because their natural ecological
niche is also thermally stable. On the basis of these observed variations, some

mammalian species have been described as 'primitive' thermoregulators, which might be taken to imply that the finer control of the 'more perfect' homeotherm has evolved independently in different evolutionary lines subsequent to the main radiation of the mammals some 70 million years ago.

Comparative observations give rise to three questions: 1) if the finer control of homeothermy has evolved independently in different mammalian lines, how safe is it to make comparisons between, say, the thermoregulation of the cat, sheep and man which belong to quite different orders? ii) Is there an evolutionary significance in the varying degrees and patterns of thermoregulation in reptiles, monotremes, marsupials, and in mammalian orders? or iii) are the species differences related more to the ecological niches in which a species is able to survive and thrive? These issues are too involved to be discussed in these brief essays, but it is to be hoped that the essay on comparative thermoregulation will render us aware of the broad biological context within which the temperature regulation of the full-time homeothermic mammal must be considered.

The essay by Professor Michel Cabanac (Lyon, France) on behavioural thermo-regulation has been placed immediately after that on comparative temperature regulation, so as to emphasize the need to consider both behavioural and autonomic components in the control of body temperature. Most of the now classical studies of thermoregulation have been concerned with the autonomic functions (i.e. those that operate, or can operate, independently of the cortex and conscious awareness or will). The undergraduate may be excused if he asserts that we shiver when we are very cold and sweat when we are very hot, and that in between extremes we rely on variations in peripheral vasomotor tone to maintain a balance between heat production and heat loss. However, Professor K.E. Cooper once told me that there is no substantial evidence of an increase in thermal lability in man following near total sympathectomy, and the reason for this may be that the fine control is achieved by behavioural means: i.e. in the choice of variations in clothing and bedding and the selection of the indoor macro-climate.

Thermoregulating reptiles depend almost entirely on behavioural means to regulate heat uptake and loss so that body temperature is held within a narrow zone of preference. With the evolution of higher metabolic rates, the birds and mammals seem to have evolved autonomic means of controlling the rate of heat loss, but this would seem to be additional to rather than in place of behavioural control. These two components of thermoregulation could be complementary but independent functions, or they could be fully integrated functions. Recent evidence indicates that in some species at least induced variations in hypothalamic temperature can influence both behavioural and autonomic thermoregulatory functions. Thus it is possible that there is some degree of integral control over these two groups of responses to thermal disturbance. The thought with which I hope the reader is left is that behavioural and autonomic functions must both be taken into account in our attempts to understand the nature of physiological temperature regulation.

Whether a control system can be taken to pieces and examined piece by piece, or whether it is sealed within a "black-box", the first examination will consist of an analysis of the relations between disturbance (or stimulus) and response. If a reproducible pattern emerges, these relations can be given a mathematical description, which can then be compared with those which describe established control processes employed by engineers. Thus it may be possible to say that the system under investigation acts as if it contains a particular control feature. It may, indeed, actually have this feature incorporated into its design, but similar functions do not necessarily depend on similar forms of engineering. Clearly the brain does not really contain control units which in anyway resemble the mechanical or electrical components employed in engineered control systems. Virtually all that the physiological controller contains is a complex pattern of

neurons and synapses which may perform equivalent functions, but in quite different ways. Thus physical and physiological controllers may have nothing in common
except a similarity in the equations which describe the relations between the input from disturbance sensors and the output to regulatory effectors.

In the essay by Dr Duncan Mitchell and his colleagues (Johannesburg, South
Africa) mathematical and engineering models which express the disturbance/response
relations, and describe analogous systems of control are discussed. The analyses
of the functions of the physiological controller indicate that the system contains
some biological equivalent to a set-point or reference signal with which a signal
derived from a sensor of the controlled-variable (core temperature?) is compared
and that the output to the thermoregulatory effectors is qualitatively and quantitatively related to the direction and magnitude of the difference between these
two signals.

To accept that the physiological controller really does behave just like this,
we would need to have some experimental evidence of neuronal functions that could
operate to this end. Electrophysiological techniques involving the precise place-
ment of fine wire electrodes deep in brain tissues, permit the monitoring of the
electrical activities of single neurons. In the preoptic region of the anterior
hypothalamus, known from classical abalation and stimulation experiments to be
concerned im the regulation of body temperature, units have been found which respond to changes in local, peripheral or spinal temperature. Some of the recorded
activity/temperature relations suggest a direct effect of temperature on the unit,
while others have been interpreted as evidence of an indirect influence of temperature. That these units are involved in the control of body temperature is an
unproven inference but there can really be little doubt that these nerve cells are
part of the thermoregulatory machinery even if their precise roles are not those
that we now thin they are. This field of research is discussed in essay by Dr
J.S. Eisenman (New York, U.S.A.) who was one of its pioneers and continues to be
one of its leaders.

The possible roles of putative synaptic transmitter substances in the central
control of body temperature is discussed by Dr Richard Hellon (London, England).
That synapses, and therefore transmitter substances, are involved in the control
of body temperature must be regarded as a basic premise in a consideration of the
nature of the mechanism, but an attempt to use synaptic interference as a means of
elucidating the nature of the thermoregulatory mechanism would not seem to afford
much hope of success. With so few candidates as central nervous transmitter
substances, and so many synapses involved in so many different functions, it might
be supposed that even the very local application of a putative transmitter substance, or other substances which affect synaptic activity, might elicit something
akin to a mass reaction, and that the consequent confused effects on thermoregulatory effector functions would be unanalysable. Contrary to this gloomy prediction, a great number of clear cut, repeatable, and apparently meaningful observations have been reported. As with the unit activity studies, there is plenty
of scope for alternative interpretations and imaginative extrapolations.

The evidence strongly indicates that noradrenaline, 5-hydroxytryptamine and
acetylcholine play roles in thermoregulation that can be distinguished from the
many other roles that these substances undoubtedly play in the central nervous
system. However, there is evidence of quite large species variations in the
functions of these different substances. For those who see an evolutionary pro-
gression in central control of body temperature which may have preceded the
emergence of the mammals, the apparently differing roles of these transmitter
substances in different mammalian species is surprising and perplexing. It is
possible that there is some other explanation for these alleged species differences, but at present their significance cannot be assessed. Recent studies have
suggested the involvement of prostaglandins in thermoregulation but no differences

in the effects of these substances were seen in species which have responded quite differently to the monoamines. Possibly the prostaglandins have a more direct and fundamental effect on thermoregulation than do the more generally accepted trans- mitter candidates.

The prerogative of the preoptic hypothalamus as the location of the deep body thermosensors became almost a dogma, and since peripheral temperature varies con- siderably while core temperature is held relatively constant, it has been assumed by many investigators that the controlled variable is the core temperature and that this is detected by temperature sensors in the hypothalamic region. The role of the peripheral temperature sensors would seem to be that of providing an early warning of any change in the thermal relations between organism and environment, and by some interactions with the signals from the central sensors, the peripheral thermal stimuli facilitate appropriate thermoregulatory responses before there is a disturbance, or gross disturbance, of the core temperature. There is a vast literature dealing with the roles of peripheral and hypothalamic temperature sen- sitivity in the control of body temperature in different circumstances and in different species. Conclusive evidence of the monopoly of the hypothalamic sen- sors of core temperature has never been given, and now there is evidence that in some species, at least, temperature sensitive structures in the spinal cord and perhaps elsewhere in deep tissues contribute theremal information to the thermo- regulatory centres in the hypothalamus. This extrahypothalamic deep-body thermo- sensitivity is discussed by Dr F.W. Klussman*from Professor Rudolph Thauer's laboratory, which has produced the main body of evidence. There is now growing support for the view that the controlled variable is not a single core temper- ature but mean body temperature, or something close to it. This might be taken to imply the presence of thermosensitive structures in many parts of the body, though perhaps in quite low concentrations.

Whether it is yet possible to transfer our thoughts on the nature of thermo- regulation from engineering to neuronal concepts is the question which I have tackled in my essay on neuronal models. The task of the neuronal model maker is to consider how a biological system composed essentially of neurons and synapses can perform functions which have distinct similarities to those of mechanical and electronic systems created by engineers to control aspects of our technolog- ical environment.

The neuronal models which are discussed in my essay are neither analogues, nor representations of established neural pathways and events. Essentially, they are speculative syntheses based on the rather sparse observations of the relations between sensors and effectors, and of the activities and interactions of neurons that may be involved in the central control of body temperature. A point of considerable interest and possible significance is that models derived in three different laboratories on the basis of entirely different kinds of evidence have much in common. All suggest that in its simplest form, the interphase between temperature sensors and thermoregulatory effectors can be thought of as being composed of two main nerual pathways: from warm sensors to heat loss effectors, and from cold sensors to heat production effectors; with crossed inhibitory path- ways between them. One does not have to look very long at the problems and the evidence to appreciate the absurd naivety of all the neuronal models so far pro- posed, but this is equally true of similar attempts to understand every other function of the brain in terms of neuronal pathways and networks.

I asked Dr H.T. Hammel (San Diego, California, U.S.A.) to give us his op- inion about the reality of the setpoint concept, because his mastery of both the physics and the physiology of temperature regulation and his ideas about an adjustable set-point in the physiological temperature regulation placed him in a better position than most to grapple with this contentious problem.

* and Dr Pierau

Whether the biological controller actually uses an error signal derived from a comparison between a stable set-point signal and a signal from a controlled variable remains to be seen, but since we can describe what we do not fully understand only in terms of things we understand somewhat better, we must continue to speak of the set-point of temperature regulation even though we are unsure exactly what we mean by the term in the biological context.

Body temperature rises during exercise, but whether this is due simply to a temporary imbalance between the rate of heat loss and the higher rate of heat production, or to an upward shift in the set-point of body temperature during exercise has been intensively studied and discussed. Originally this issue was to be debated by Dr D. McK. Kerslake of Farnborough, England, but an illness compelled his withdrawal and Dr Jan Snellen (New Foundland, Canada) bravely accepted the challenge at short notice. The interesting point of Snellen's analysis is that he supports the proposition that there is a multiple input from temperature sensors in many parts of the body. The shift in the balance between heat production and heat loss, and thus in core temperature, could be attributed simply to a shift in the balance of the signals from temperature sensors in different parts of the body.

An upward shift in set-point during fever appears to be a very apt description, for during the period of a steady elevated temperature during fever, a subject will respond to heat stress or cold stress or to exercise in much the same way as a normal subject. The only difference seems to be in the level of control. All the evidence indicates that this upward shift in the set-point is due to an effect of endogenous pyrogen in cells in the preoptic region of the hypothalamus. This aspect of the control of body temperature is discuss by Professor K.E. Cooper who studied fever and other aspects of temperature regulation for many years, until recently at the Radcliffe Infirmary, Oxford, and now at Calgary, Canada.

A case could be made for the inclusion of essays on the ontogenesis and phylogenesis of mammalian homeothermy, and on the nature of hibernation: whether the set-point is switched off, or switched down. It is always difficult to know where to stop. Some readers will consider that already we have cast our nets too widely, others not widely enough. However, the volume ends with a summarizing essay by Dr J.D. Hardy (New Haven, U.S.A.) who concentrates on the main theme of the essays - the validity of the models without which we cannot even think of the complexities of a biological control system, let alone begin to understand it. His essay is a clear exposition based on both experience and vision, and is a masterly example of the lucid contributions he has been making to almost every corner of the field for nearly forty years.

John Bligh

A.R.C. Institute of Animal Physiology,
Babraham, Cambridge, England.

PRIMITIVE MAMMALS AND PATTERNS IN THE
EVOLUTION OF THERMOREGULATION

TERENCE J. DAWSON
School of Zoology, University of New South Wales,
Kensington, N.S.W.   Australia

Abstract: Recent evidence has indicated that the various groups of
mammals are more closely related than has been usually considered.
As a consequence of this a review has been made of the thermoregulatory abilities of the more primitive mammals such as monotremes, marsupials, insectivores and edentates, to see if these
show a distinct pattern which could yield insight into the evolution of thermoregulation. In the more primitive groups there is
an overall tendency toward low body temperatures and low levels of
metabolism and the suggestion has been made that the higher body
temperature of advanced eutherians is a relatively recent development. No clear pattern has emerged with respect to the modes of
evaporative heat loss in primitive groups. The possibility of an
independent evolution of panting and sweating in the marsupials
and placentals is discussed.

1.   INTRODUCTION

        This symposium is concerned with the control of body tem-
perature and looking down the programme I see a range of erudite
studies on the mechanisms involved in the regulation of body temperature. In recent years I have worked on the temperature regulation
of marsupials and monotremes, the so-called primitive mammals, and
consequently I would like to be able to present to you a paper describing some nice primitive control system or model thereof, but unfortunately I can't. There are two reasons for this:

        (1) The state of knowledge is only now reaching the level
where we have a reasonably accurate assessment of the thermoregulatory abilities and characteristics of some of these primitive animals.
It serves no purpose to place thermodes into the hypothalamus if you
don't know what are the animals' normal thermoregulatory responses.

        (2) It might not be possible to show a nice primitive con-
trol mechanism simply because the control mechanism may not be primitive. While a marsupial, such as a kangaroo, may have a relatively
low level of metabolism and body temperature, which may be suggestive
of a primitive mammalian condition, this does not mean that it cannot
regulate its body temperature at its chosen level as efficiently as
the advanced or higher forms. It may be the furnace that is primitive not the thermostat.

        What I would like to do in this paper is to have a look at
the comparative physiology of thermoregulation in primitive mammals
and see what this can tell about the evolution of homeothermy and
therefore give some indication where one should start looking if one
wants to examine some of the characteristics of an early mammalian
control system. The problem with the comparative approach to this
sort of question is the very real difficulty associated with knowing
what are the relationships of various animals to each other, in both
the past and the present. So initially I would like to spend a

little time reviewing current ideas on the phylogenetic relationships
of primitive mammals.

2.   PHYLOGENY OF PRIMITIVE MAMMALS

        The usual starting point in a discussion of the evolution
of homeothermy in mammals is with the reptiles, particularly the more
advanced lizards of the order Squamata. Many studies have shown that
lizards are able to control their body temperature by behavioural
means within narrow limits (see Templeton 1970 for review). Studies
have also been carried out into the neural control of this type of
body temperature regulation; Hammel *et al* (1967) have implicated the
reptilian preoptic region in behavioural regulation and Cabanac *et al*
(1967) working with the same lizard (*Tiliqua scincoides*) have demonstrated the presence of both warm and cold sensitive neurons in the
lizard hypothalamus. These authors preferred not to come to a firm
conclusion regarding the existence of a link between these neurons
and thermoregulation, yet they suggested that they may be the roots
of the later evolved physiological hypothalamic thermostat. The problem with extrapolating these studies to what happened in the evolution of thermoregulation in primitive mammals is that the fossil
evidence shows that the ancestry of the mammals separated from the
ancestral cotylosaurs and the line that gave rise to the living reptiles very soon after the origin of the class Reptilia some 300 million years ago (Hopson 1969). This fact is significant in that it
indicates that many of the features characteristic of living reptiles
were not present in the early reptilian ancestors of the mammals, and
therefore, modern reptiles cannot be considered to represent an
"evolutionary stage", preceding mammals. This also explains the pe-
culiar fact that some nineteenth century anatomists thought it easier
to derive mammals directly from primitive amphibians than from reptiles as they were then known (Hopson 1969).

        The history of the line which gave rise to the mammals, the
synapsid reptiles, can be divided into a series of adaptive radiations, each of which was monophyletically derived from the preceding
one. The basic theme was the evolution of jaw attachments and teeth
so that it was possible to masticate the food. This preparation of
food in the oral cavity before passing it down the digestive tract is
very important in providing for a faster digestion and the rapid de-
livery of energy to the body. Needless to say the necessary bio-
chemical changes would have to go hand in hand with these develop-
ments.

        Recent information (Hopson 1969, Hopson & Crompton 1969,
Barghusen & Hopson 1970, and Parrington 1971) indicates that all the
Mesozoic primitive mammals were much more closely related to one another than was previously thought (Figure 1). In contrast to theories of the polyphyletic origin of these mammals from different ancestral groups of therapsid reptiles (Kermack 1967), the evidence now
indicates that mammals were derived from a cynodont ancestor, probably within the family Galesauridae in the late Triassic over 200
million years ago.

        Unfortunately there are no mammal-like reptiles such as the
cynodonts still around for study by comparative physiologists, but
there are survivors of the early radiation of non-therian mammals.
These are the monotremes of Australia, the echidnas or spiny anteaters (*Tachyglossus* and *Zaglossus*) and the platypus (*Ornithorhynchus*).

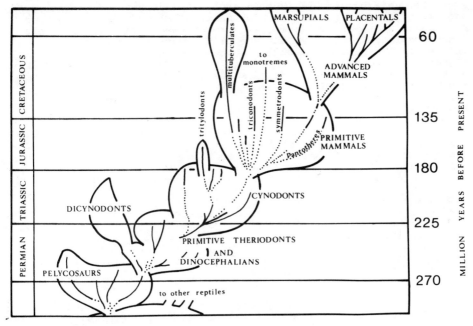

Fig. 1.   An outline of the successive radiations of synapsid and mammalian groups.   After Hopson 1969.

The monotremes have retained many of the features presumed to characterize the earliest mammals, and are therefore uniquely qualified among living tetrapod-vertebrates to yield information about the physiology and anatomy of early mammals.   There apparently was, however, a very early separation (about 200 million years ago) of the stock leading to the living marsupials and placentals, the therians, from that which gave rise to monotremes (Hopson 1969).Although the monotremes possess many primitive features they have clearly attained a mammalian level of organization.   This strongly suggests that the common ancestor of the monotremes and therians was also mammalian in a majority of essential features e.g. hair, lungs, diaphragm, heart, kidneys etc.

        The last great radiation has been the radiation of the advanced or therian mammals which include the marsupials or Metatheria and the placentals or Eutheria.   Lillegraven (1969) in his review of the marsupial – placental dichotomy in mammalian evolution suggests that the marsupials and placentals have been distinct for a long time, perhaps since the earliest Cretaceous about 130 million years ago.   The other interesting point to come out of this review was the conclusion that the common ancestor of both groups was probably much more "metatherian", i.e. marsupial like, than eutherian than has been considered in recent years.   So while not on the direct line of descent, the more primitive marsupials (such as the smaller opossums and dasyurids) in their mode of life and many structural features, may give a picture of the Mesozoic forms from which the Tertiary mammals have come.

        There are five super families of living marsupials, three of them Australian;   the Dasyuroidea - carnivorous mice and cats etc.,

Perameloidea - bandicoots, Phalangeroidea - possums and kangaroos,
and two of them American, the Didelphoidea and Caenolestoidea. The
Australian and American forms have probably been genetically isolated
since the late Cretaceous, but one of the more interesting pieces of
information to come up recently is that the American didelphids
appear to be more closely related to the Australian forms than either
of these groups is to the caenolestids (Hayman *et al* 1971). These
mouse-like caenolestids are apparently a very early off-shoot of the
marsupial line, but unfortunately nothing is known about their physiology.

When one comes to the "higher", or eutherian mammals and
starts hunting around the phylogenetic tree for some primitive animals one quickly finds some more interesting information. The eutherian mammals actually comprise two lines which have been genetically
distinct for some 75 million years. These groups have been based on
the primitive leptictid insectivores and the palaeoryctid insectivores
(McKenna 1969). The leptictid group, according to McKenna (1969),
gave rise to the tree shrews, primates, rodents and perhaps the lago-
morphs (rabbits and hares). The palaeoryctids gave rise to most of
the insectivores and the carnivore-ungulate group. Thus it would be
well for workers to keep this in mind when talking of higher mammalian characteristics since a great span of geological time for diver-
gent and convergent evolution has been available since these basic
groups separated.

In comparative physiology there is always a danger in
extrapolating from one animal or a small group of animals. Selection
acting on similar structures inherited from a common ancestral stock
may independently produce similar solutions to similar functional
problems. Therefore one should look for overall patterns from several groups if one wishes to determine the phylogeny of a particular
function. Therefore, to look at temperature regulation, I have
gathered together the various eutherians that are considered primitive or at least close to the ancestral type of their group (Table 1)
and examined what is known about their temperature regulation together with what is currently known about that of marsupials and
monotremes and endeavoured to see what overall patterns emerge.

Table 1

Eutherian mammals usually considered primitive

1. Leptictid Group.
        Order Insectivora
            Family Macroscelididae   Elephant shrews

        Order Dermoptera
            Family Cynocephalidae   Flying lemurs

        Order Primates
            Family Tupaiidae   Tree shrews
            Family Lemuridae   Lemurs
            Family Indridae   Indrisoid lemurs
            Family Lorisidae   Lorises and galagos

        Order Rodentia
            Family Aplodontidae   Mountain beaver

2.   Palaeoryctoid Group.
        Order Insectivora
            Family Tenrecidae   Tenrecs
            Family Solenodontidae   Solenodons
            Family Erinaceidae   Hedgehogs
            Family Soricidae   Shrews

        Order Tubulindentata
            Family Orycteropodidae   Aardvark

        Order Hyracoidae
            Family Procaviidae   Hyraxes

3.   Affinities Uncertain.
        Order Edentata
            Family Myrmecophagidae   Anteaters
            Family Bradypodidae   Tree sloths
            Family Dasypodidae   Armadillos

        Order Pholidota
            Family Manidae   Pangolins

        In this review I have omitted discussion of the thermo-
regulatory characteristics of the bats because of the great diversity
of thermoregulatory patterns in the order Chiroptera. While bats, in
general, have lower metabolic rates than would be predicted for
eutherian mammals from the equation of Kleiber (1961) (Henshaw 1970),
an adequate discussion of the possible significance of these patterns
in the evolution of homeothermy in mammals is beyond the scope of
this review.

3.   THERMOREGULATORY PATTERNS

Body temperature.
        When one examines the values that have been obtained for
the body temperatures of various primitive mammals, the monotremes,
marsupials and less advanced eutherians, the general pattern that
becomes apparent is a tendency toward low body temperatures (some
representative values are given in Table 2). When it comes to low
body temperatures the monotremes are the extreme case with all three
genera having resting deep body temperatures around 30°C in or near
their thermo-neutral range. The marsupials also tend to have body
temperatures below those of the advanced eutherians, the general
range for both the Australian and American forms being about 35°C.
Some of the smaller dasyurids may have resting body temperatures
which are generally below this level. Unfortunately nothing is known
about the caenolestids.

        The primitive eutherians also show the same pattern of a
low body temperature (Table 2). This is not always the case since
the shrews of the genus *Sorex* have quite a high body temperature but
the other more primitive insectivores, such as the tenrec and hedgehog, have low temperatures. The low body temperature of the edentates, sloths, armadillos etc. has been well known for some time. So
then the overall pattern indicates a generally lower body temperature
and it would seem that the ancestors of the modern mammals were probably similarly endowed. A high body temperature is probably a

Table 2

Body temperatures of some primitive mammals*

Animal	Body temperature (°C)	Reference
Order Monotremata		
Ornithorhynchus anatinus (platypus)	(30.0-32.7)	59, 65
Tachyglossus aculeatus (echidna)	(28.6-31.6)	48, 69
Zaglossus sp. (long-nosed echidna)	29.0 (26.2-31.8)	67
Order Marsupialia		
Didelphis marsupialis (opossum)	35.0 (34.0-36.5)	52, 57
Metachirus nudicaudatus (brown opossum)	33.8 (32.4-36.4)	52
Chironectes panamensis (water opossum)	35.2	11
Sminthopsis crassicaudata (fat-tailed marsupial mouse)	33.8 (32.7-35.2)	21
Antechinus stuartii (brown marsupial mouse)	34.4 (33.5-35.7)	21
Satanellus hallucatus (native "cat")	(34.0-35.2)	55
Sarcophilus harrisii (tasmanian devil)	(36.0-36.1)	47
Perameles nasuta (long-nosed bandicoot)	36.1 (35.0-36.8)	21
Isoodon macrourus (short-nosed bandicoot)	34.7 (33.7-35.8)	21
Petaurus breviceps (sugar glider)	36.4 (35.6-37.7)	21
Trichosurus vulpecula (brush-tailed possum)	36.2 (35.4-36.7)	16, 21
Macropus eugenii (tammar wallaby)	36.4 (35.6-36.8)	21
Megaleia rufa (red kangaroo)	35.9 (35.4-36.8)	21
Order Insectivora		
Tenrec ecaudatus (tenrec)	33.0 (32.5-34.0)	35
Hemiechinus auritus (long-eared desert hedgehog)	(33.4-36.4)	24
Paraechinus aethiopicus (desert hedgehog)	(31.2-36.2)	24
Erinaceus europaeus (eurasian hedgehog)	35.6 (34.8-36.4)	54
Sorex cinereus (masked shrew)	38.8 (36.0-40.5)	58
Sorex palustris (water shrew)	39.7	14
Order Pholidota		
Manis tricuspis (pangolin)	(32.2-35.2)	24
Order Edentata		
Myrmecophaga jubata (giant anteater)	(32.0-34.0)	72
Tamandua tetradactyla (collared anteater)	33.5 (32.0-35.0)	25
Bradypus griseus (three-toed sloth)	(32.2-34.1)	25
Choloepus hoffmanni (two-toed sloth)	(34.2-35.8)	25
Dasypus novemcinctus (nine-banded armadillo)	(33.9-34.5)	42
Order Primates		
Nycticebus coucang (slow loris)	34.9	66

Single values are means and values in parentheses show ranges.

* Where possible body temperatures were obtained for resting animals in their thermo-neutral zone.

recent (less than 70 million years ago) acquisition.

Thermogenesis.

Why a low regulated body temperature? Since all these animals were supposedly endothermic it is therefore necessary to look at
metabolism. The only valid initial basis for comparison of metabolism is basal metabolic rate (B.M.R.) or standard metabolism. This
is the minimal level of metabolism and it is attained in thermo-
neutral surroundings in a post-absorptive state and during minimal
physical activity. It is thought to represent the heat expenditure
associated with the maintenance of those functions of the body necessary to maintain life awareness. Under conditions of basal metabolism the level of heat production is actually in excess of the
requirement for the maintenance of body temperature. In reality it
reflects simply the fundamental level of metabolic organisation and
activity of the animal. Other levels of metabolism, such as the
maximal metabolism, are generally related to this base level (Janský
1965). This is not entirely true for the non-fasting metabolic rate,
since the specific dynamic effect of feed is related to the type as
well as the amount of feed ingested (Kleiber 1961).

Table 3

Basal metabolism of mammals and reptiles*

	Reptiles	Mammals		
	Lizards	Monotremes	Marsupials	Placentals
Approximate body temperature (°C)	30	30	35.5	38.0
Basal metabolic rate Kcal/kg$^{3/4}$ day	7.5	34	49	69
Corrected (38°C) BMR Kcal/kg$^{3/4}$ day	19.5	62	62	69

* derived from Dawson and Hulbert (1970).

Information concerning the metabolism of primitive mammals
tends to be sparse and only recently have reliable data for monotremes and marsupials become available. Table 3 shows that both the
monotremes and marsupials have basal metabolic rates which are considerably below those found for most eutherian mammals (Kleiber 1961).
Values reported for the monotreme *Tachyglossus aculeatus* range from
37% (Augee and Ealey 1968) to 49% (Schmidt-Nielsen *et al* 1966) of the
predicted eutherian level. There is now good agreement that the marsupials have basal metabolic rates which are approximately 70% of the
predicted eutherian rate. Dawson and Hulbert (1969 & 1970) have
shown this to be true for a wide range of Australian marsupials and
MacMillen and Nelson (1969), have reported similar levels for 12
species from the family Dasyuridae. Other workers, Bartholomew and
Hudson (1962) and Arnold and Shield (1970) have obtained similar data
from single species. The little available information about the
metabolic rates of American didelphids indicates that they may be

similar to the Australian super families (Morrison and McNab 1962,
T.J. Dawson, E.C. Crawford and K. Schmidt-Nielsen unpublished observation cited by Dawson and Hulbert 1970). Unfortunately nothing is
known about the South American caenolestids.

Among the eutherian groups that are generally considered to
be primitive there appears to be a tendency toward low basal metabolic rates. The low metabolic rate of various species in the order
Edentata has been known for some time. The sloths were examined by
Irving *et al* (1942) who found that the two toed sloth *Choloepus* and
the three toed sloth *Bradypus* have basal metabolic rates which are
well below the predicted eutherian levels. The armadillos have been
studied by a number of workers and the reported basal metabolic rates
are, in most cases, similar to that found for *Dasypus novemcinctus
mexicanus* by Johansen (1961) which was 57% of the predicted value.
Enger (1957) has shown that the low level of metabolism also extends
to the anteaters of the family Myrmecophagidae.

There has been some controversy concerning the metabolic
status of the various groups in the Insectivora. Recent work carried
out under well controlled conditions has demonstrated that the tenrec
(*Tenrec ecaudatus*) and the hedgehog (*Erinaceus europaeus*) have low
basal metabolisms. Hildwein (1970) and Hildwein and Malan (1970)
have shown that basal rate of metabolism of the tenrec and hedgehog
respectively is 44-64% and 71% of the values predicted from Kleibers'
equation (Kleiber 1961). Shrews of the family Soricidae have been
reported as having a very high relative B.M.R. (Pearson 1948;
Morrison 1948; Morrison *et al* 1959). Shrews are small active animals and measurement of an accurate B.M.R. would appear to be difficult. Most of these early measurements were also made on non-fasting
animals at air temperatures which were probably well outside the
animals' fasting thermo-neutral zone. Pearson (1948) carried out his
investigations at an air temperature of 24°C; experience with other
small homeotherms such as the hummingbirds (Lasiewski 1963) and
dasyurid marsupials (MacMillen and Nelson 1969, Dawson and Hulbert
1970) would suggest that this temperature was considerably below the
lower critical temperature of at least the smaller species of shrews.
A study by Hawkins *et al* (1960) of the metabolism of some British
shrews at higher air temperatures, 28-30°C, and under more controlled
conditions (still non-fasting though) indicated that the resting
metabolism of these shrews was not higher than was to be expected on
the basis of size alone.

A most interesting study has been carried out by Redmond
and Layne (1958). These workers examined the metabolic rates of the
tissues of several species of shrew and found that in general the
levels were below that predicted from studies on other eutherians.
Values also obtained from the tissues of one mole (*Scalopus aquaticus*)
were also low and Redmond and Layne suggested that the insectivores
may have inherently low tissue metabolic rates. Obviously the metabolic status of shrews needs further investigation but it is possible
that these insectivores may actually have a relatively low B.M.R.

Little is known of the metabolism of other eutherians commonly regarded as being primitive. The hyrax *Heterohyrax brucei* has
a relatively low B.M.R. (Bartholomew and Rainy 1971) and values derived from the work of Yousef *et al* (1971) indicate that the non-
fasting metabolism of a primitive primate, the tree shrew *Tupaia
chinesis*, is similar to the predicted basal level. Consequently it
is likely that the B.M.R. (post-absorbtive metabolism) would be below the predicted level.

It would appear then that a low level of metabolic organization may be characteristic of primitive groups and therefore
characteristic of the earliest mammals. One would assume also that
the capacity for increase, i.e. the metabolic scope for temperature
regulation, would be related to the base rate as it is in other
species (Giaja 1938). How, then, does this fit in with the regulation of body temperature, since I have stated that when the B.M.R.
is taken there is an excess of heat production over that required for
temperature regulation? Perhaps the best way is to start with the
familiar heat production-environmental temperature curve (Figure 2).
This is a gross over-simplification of what in reality occurs but it
is a reasonable starting point.

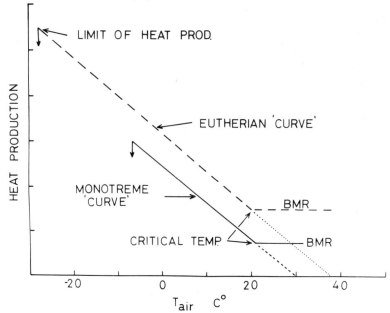

Fig. 2. Comparison of eutherian and monotreme heat production-air
temperature "curves". Based on data from Schmidt-Nielsen *et al* 1966.

Below the critical temperature the basal heat production is
not sufficient to maintain body temperature. The animal must then
produce extra heat. The amount of heat required will be dependent on
the total insulation and is proportional to the difference between
air temperature and body temperature (Newtons "law" of cooling).
Burton and Edholm (1955) give an excellent explanation of why this
empirical "law" holds over   a great range of environmental conditions. Now what happens if the level of metabolism is low as in
the monotreme *Tachyglossus aculeatus* where the base level of metabolism is below ½ that of a similar sized advanced eutherian? If this
animal was to maintain the same body temperature as the eutherian and
presumably had the same total body insulation (core to environment),
then several things become apparent.   (1) The monotreme would have to
increase heat production just for the maintenance of body temperature,
at ambient temperatures much higher than would the eutherian (a waste-
ful use of energy).   (2) If there was a limit to the metabolic increase, say 4 times the basal value (Giaja 1938), then the monotreme
would run out of its metabolic range at much higher ambient temperature than would a eutherian. If, however, the monotremes settled on

a lower body temperature, 29-30°C, then for no extra trouble they would extend their lower activity range by 8-9°C. If environmental temperatures fall below this range the options open are hibernation, or tolerance of a lower body temperature for a while.

The marsupials are in a similar position; the metabolism is low, approximately 70% of the predicted eutherian level. They also have relatively low body temperatures but not as low as monotremes, so apparently there is more than one way of coping with the problem. The slope of the thermostatic line is dependent on total body insulation (body core to environment), the greater this total insulation, the lower the slope and consequently the smaller the amount of heat required to maintain body temperature at low environmental temperatures. So far it appears that, as shown in Figure 3, the total body insulation of marsupials is much greater than that found in eutherians of comparable size and habitat (Dawson *et al* 1969). The insulation is not due to a high fur insulation since fur insulation is relatively low (Dawson and Brown 1970). Dawson *et al* (1969) have suggested that the tissue insulation is high and this is due to a low peripheral blood flow which is related to the low level of metabolic activity. In some of the other primitive mammals specialized mechanisms are involved in this process. Scholander and Krog (1957) have shown that sloths have arteriovenous bundles or rete in their limbs, which function to reduce heat loss from the body core, thus increasing total body insulation. Other animals that have similar structures are the edentate anteaters, the pangolin and lorisine lemurs (Barnett *et al* 1958).

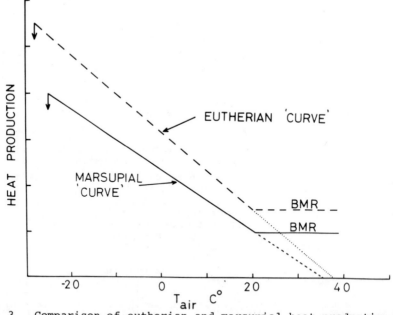

Fig. 3. Comparison of eutherian and marsupial heat production-air temperature "curves". Based on data from Dawson *et al* 1969.

From the foregoing information it would seem that the attaining of a high and stable body temperature in the course of mammalian evolution, and its consequent speeding up of body processes, was initially limited by the fundamental metabolic capability of the early mammals and their need to live in varying environments. One

mechanism employed by some of the living primitive species to overcome this limitation and increase their body temperature and environmental temperature range has been an increase in maximum body insulation.   In this regard, however, a large increase solely in fur
insulation would not be satisfactory since this would cause problems
with heat loss during exercise and in hot environments.   One point to
consider is that a relatively stable body temperature is apparently
preferred to a widely varying one, even if the stability is at a
lower level.   The body temperatures of some marsupials, at least, do
not vary significantly in the thermoneutral zone even though excess
heat is available (Dawson *et al* 1969).

        Another finding that has come out of studies of comparative
metabolism is that homeothermy seems to be associated with a major
jump in metabolic activity over that in poikilothermic animals.   If
the basal metabolisms are corrected to a specific body temperature
all the homeotherms come out at a similar level, which is very much
above that of the advanced poikilotherms (Dawson and Hulbert 1970).
In this regard the recent work of Ismail-Beigi and Edelman (1970) on
the mechanism of thyroid calorigenesis is of particular interest.
These workers have demonstrated that thyroid calorigenesis is mediated by stimulation of the active transport of sodium in cell membranes,
In mammals, therefore, cell membranes are acting, in part, as a specific heat producing system which is not available to the poikilothermic or ectothermic vertebrates.

Heat Dissipation.

        The other aspect of the control of body temperature which
has not yet been discussed is the regulation of heat loss.   Below the
lower critical temperature insulation is usually maximal in order to
retard heat loss.   Above this critical temperature an excess of heat
is produced over that required for the maintenance of body temperature and this heat must be dissipated.   The major avenues of heat
loss are physical or dry heat loss (radiation, conduction and convection) and evaporative heat loss which is dependent on the latent
heat of vaporization of water.

Physical Heat Loss

        The heat loss from an animal by radiation, conduction and
convection is dependent on the temperature gradient and the total insulation.   The variation of total insulation by varying the peripheral blood flow is probably a very ancient mechanism.   Contrary to
the suggestion of Martin (1902) the monotremes do appear ·to have this
ability (Schmidt-Nielsen *et al* 1966).   This ability is also widely
distributed in the various groups of reptiles (Templeton 1970).

Evaporative Heat Loss

        In the absence of a sufficient gradient for adequate dry
heat loss, heat can only be dissipated by the evaporation of water.
There are three possible methods of evaporative heat loss, licking,
panting, and sweating, and all are used by various groups of mammals.

        Licking.
        Licking has always been a problem in thermoregulation be-
cause it is so difficult to gauge its usefulness.   It has been regarded as the most primitive form of active evaporation and has commonly
been described as the main form of evaporative heat loss in marsupials (Higginbotham and Koon 1955; Bartholomew 1956).   Recent work
has shown that while many marsupials do lick in response to high

environmental temperatures (others do not) this is of minor importance, panting being the major avenue of active evaporative heat loss
(Dawson 1969;   Dawson *et al* 1969).

The reason why licking has been given so much prominence
over panting in marsupial temperature regulation is probably due to
the fact that it is so obvious, whereas water loss via the respiratory tract and particularly from efficient sweating is not obvious.
Bartholomew (1956) working with the quokka (*Setonix brachyurus*), a
small macropodid, observed copious licking and stated that it seemed
certain that licking of the paws and tail was primarily responsible
for the quokka's heat regulation at an environmental temperature of
40°C.   Panting he assumes is not important to the quokka because,
"Although their respiratory rate may accelerate to as much as 200/min,
the breathing of quokkas subject to heat stress never approaches in
vigor the panting of dogs".   It is difficult to see how what must be
about a fifteenfold increase in respiration rate could be of minor
importance particularly since the maximum panting frequency in the
dog is only about 300 breaths/min (Crawford 1962).   Bentley (1960) has
since shown that the thermoregulatory capabilities of the quokka are
not significantly affected when the animal is prevented from licking.

Another piece of information which may have led workers to
attach due importance to licking as a means of temperature regulation
was the report by Higginbotham and Koon (1955) that the evaporation
of saliva from the body surface appeared to be an indispensable heat-
dissipating mechanism in the opossum *Didelphis marsupialis*.   This
conclusion was drawn from a sole experiment in which a single opossum
under sodium pentobarbital anaesthesia did not regulate as well as
several non-anaesthetized animals.   Higginbotham and Koon attributed
this to the lack of licking in the anaesthetized animal.   The dele-
terious effects of anaesthetic on temperature regulation are well
known (Bligh 1966) and this result should be considered as being very
suspect.   Higginbotham and Koon do report a marked increase in panting in both conscious and anaesthetized opossums.

Part of the problem in determining the importance of licking in heat dissipation comes from the fact that many workers do not
realize that open mouth resonant frequency panting, as occurs in
dogs, is not the usual form of panting response.   Most mammals which
utilize the respiratory system for active evaporative heat loss
"pant" through the nose and also show a graded respiratory frequency
response.   Failure to appreciate this point can lead to statements
like those of McManus (1971), who reported that the main response to
high temperature of the opossum is to increase evaporative heat loss
by saliva spreading even though he comments "As ambient temperature
approached and rose above 40°C, the licking behaviour often was
interupted, the opossums lying on their sides breathing deeply and
rapidly;   actual panting was not observed"........."Above T_a 41°C,
individuals appeared to be under severe temperature stress and spent
more of their time on their sides breathing heavily".   Roberts *et al*
(1969) have recently shown that radio frequency warming of the medial
preoptic areas and anterior hypothalamus of the opossum brain elicits
panting, licking, and sleep-like relaxation similar to the responses
induced by environmental heating.   Panting was more easily initiated
by these procedures than was licking.

While licking does not occur in the monotremes it is found
in various eutherian species of both the leptictid and palaeoryctid
groups.   Licking may be important in the thermoregulation of the
tenrec (*Tenrec ecaudatus*) and some rodents (Hildwein 1970;

Hainsworth 1968).   There is convincing evidence of an increased output of saliva in various groups in response to thermal stress or
hypothalamic heating (Sharp *et al* 1969;   Roberts *et al* 1969; and Antal
& Kirilčuk 1969) even when the saliva is not spread on the fur.
Licking, as a thermoregulatory response, then would appear to predate
the marsupial-eutherian dichotomy in mammalian evolution, some 130
million years ago.   Where thermoregulatory licking occurs in the
various groups of mammals it would seem that it is generally an
auxillary behavioural response, not the principal mode of evaporative
heat loss.

Panting.
Panting appears to be the most successful form of evaporative heat loss and is considered by many to be the most effective
(see Hammel 1968 and Richards 1970 for reviews).   In both these reviews there is the suggestion that mammals inherited thermal panting
from the reptiles, since some modern lizards apparently pant (Dawson
and Templeton 1966).   However, as mentioned earlier, the very early
separation of the lines which gave rise to the modern reptiles and
the mammals precludes this type of assessment.

Monotremes do not respond to heat stress by panting and in
fact the echidna *Tachyglossus aculeatus*, decreases its respiratory
rate at high air temperatures (Martin 1902).   The marsupials, in
general, use panting as their principal form of evaporative heat loss
(Robinson and Morrison 1957;   Dawson 1969, and Dawson *et al* 1969).
The pattern of the respiratory response of the tammar wallaby,
Macropus eugenii, to moderate and severe heat is so similar to that
of many eutherians (Dawson and Rose 1970) that it seems probable that
the ancestors of the marsupials and eutherians possessed the ability
to pant.   This becomes more likely when it is realized that the
armadillo (Johansen 1961), hedgehogs (Hildwein & Malan 1970, and
Scholnik & Schmidt-Nielsen personal communication)   prosimian primates (Robertshaw personal communication)   and hyraxes (Bartholomew
and Rainy 1971) all pant.   That the tenrec does not pant but relies
on licking to prevent overheating (Hildwein 1970) casts some doubt on
this suggestion, however.

Sweating.
The place of thermoregulatory sweating in the evolution of
homeothermy is enigmatic.   It would appear to be a very ancient form
of evaporative heat loss in the line leading to the mammals, since it
occurs in all three groups of mammals.   The platypus *Ornithorhynchus
anatinus* is reported to sweat (Martin 1902);   sweating also occurs in
the large kangaroos (Dawson 1971) and in the principal eutherian
lines.   However, there is a flaw in the suggestion that this indicates a phylogenetically old mechanism since general body sweating is
not found in most primitive species and the systems involved in the
regulation of sweating may be very different in various groups.
Robertshaw (at this meeting) has shown the diversity of control mechanisms in the eutherians.   Perhaps sweating is an example of basic
structures being called into operation by different groups in different ways.   Robertshaw has suggested that the efficiency of panting
is limited by body size and that many animals whose adult weight is
greater than approximately 100 kg have evolved sweating as an additional heat loss mechanism to supplement panting.   Man and the higher
primates appear to be an unusual group in this respect.

Analysis of the information available concerning evaporative heat loss from extant primitive mammals does not show any distinct overall pattern which may be used to assess the phylogenetic

relationships of the various mechanisms, licking, panting and sweating, in the evolution of homeothermy in mammals. Much more work on
the responses of monotremes, marsupials and insectivores to high
temperature will be needed before this issue is resolved. One possibility, which at this stage should not be disregarded, is that the
evolution of the various evaporative heat loss mechanisms may have
occurred independently in the various mammalian groups. These mechanisms may not have been inherited from the mammal-like reptiles
which gave rise to the earliest Mesozoic mammals. These animals were
all small insectivores; the earliest mammals were shrew-sized
(Hopson 1969). The development of endothermy was possibly successful
because it allowed these small primitive insectivorous mammals to ex-
ploit the nocturnal environment, an environment largely denied to the
ectothermic reptiles. If this was the case, active evaporative
mechanisms, except perhaps licking, may not have been present. Small
modern mammals such as the rodents and dasyurids, which fill this
niche still do not appear to pant or sweat. The earliest eutherians
and marsupials were also very small insectivorous types.

4.   CONCLUSIONS

1.   The mammals are a much more closely related group than has been
previously considered; consequently comparative studies to gain insight into the evolution of temperature regulation become much more
valid.

2.   Mammals which are usually considered to be primitive on the basis
of comparative anatomy tend to have body temperatures and metabolic
levels which are below those of advanced eutherians. The low body
temperatures are probably related to the limited levels of heat
production.

3.   Variation of heat loss by cardiovascular adjustment occurs in all
mammalian groups and is probably an ancient mechanism.

4.   Among the therian mammals, marsupials and eutherians, panting in
various forms appears to be the usual method of evaporative heat loss.
Sweating is generally used as an auxillary heat dissipating mechanism,
although in the higher primates it is of principal importance.
Except in limited cases the value of licking in thermoregulation has
usually been over-estimated, particularly in marsupials.

5.   It is possible that the various types of evaporative heat loss
may have evolved independently in the different mammalian groups.

6.   Finally, it would seem that the basic thermoregulatory control
system may be old, since all the primitive mammals can thermoregulate
very nicely if their metabolic capabilities are not exceeded. The
fact that they may become hypothermic or hibernate in some conditions
should not be used to infer that the controlling system is primitive.
The sloth *Bradypus griseus*, when pregnant, maintains an extremely
stable body temperature in conditions where the body temperature of
non-pregnant animals would vary widely (Morrison 1945). A sensitive
control system is present even if it is not always used.

Acknowledgements: I wish to thank Dr. Eleanor Russell and A.J.
Hulbert for their helpful discussions of this subject which generated
many of the ideas presented in this review.

5. REFERENCES

1. Antal, J. and V. Kirilčuk, (1969), Dynamics of polypneic sali-
    vation in a dog. Pflügers Arch. 308 : 25-35.
2. Arnold, J. and J. Shield, (1970), Oxygen consumption and body
    temperature of the Chuditch (*Dasyurus geoffroii*). J. Zool.
    160 : 391-404.
3. Augee, M.L. and E.H.M. Ealey, (1968), Torpor in the echidna,
    Tachyglossus aculeatus. J. Mammalogy. 49 : 446-454.
4. Barghusen, H.R. and J.A. Hopson, (1970), Dentary-squamosal joint
    and the origin of mammals. Science. 168 : 573-575.
5. Barnett, C.H., R.J. Harrison and J.D.W. Tomlinson, (1958),
    Variations in the venous systems of Mammals. Biol. Rev. 33 :
    442-487.
6. Bartholomew, G.A. (1956), Temperature regulation in the macropod
    marsupial, *Setonix brachyurus*. Physiol. Zool. 29 : 26-40.
7. Bartholomew, G.A. and J.W. Hudson, (1962), Hibernation, estivat-
    ion, temperature regulation, evaporative water loss, and heart
    rate of the pigmy possum, *Cercaertus nanus*. Physiol. Zool.
    35 : 94-107.
8. Bartholomew, G.A. and M. Rainy, (1971), Regulation of body tem-
    perature in the rock hyrax *Heterohyrax brucei*. J. Mammalogy
    51 : 81-95.
9. Bentley, P.J. (1960), Evaporative water loss and temperature
    regulation in the marsupial *Setonix brachyurus*. Aust. J. Exp.
    Biol. 38 : 301-306.
10. Bligh, J. (1966), The thermosensitivity of the hypothalamus and
    thermoregulation in mammals. Biol. Rev. 41 : 317-367.
11. Britten, S.W. and R.F. Kline, (1939), Emotional hyperglycemia and
    hyperthermia in tropical mammals and reptiles. Amer. J.
    Physiol. 125 : 730-734.
12. Burton, A.C. and O.G. Edholm, (1955), "Man in a cold environment".
    (Edward Arnold Ltd.) London.
13. Cabanac, M., H.T. Hammel, and J.D. Hardy, (1967), *Tiliqua
    scincoides* : Temperature sensitive units in lizard brain.
    Science, 158 : 1050-1051.
14. Calder, W.A. (1969), Temperature regulation and underwater en-
    durance of the smallest homeothermic diver, the water shrew.
    Comp. Biochem. Physiol. 30 : 1075-1082.
15. Crawford, E.C., Jr. (1962), Mechanical aspects of panting in
    dogs. J. Appl. Physiol. 17 : 249-251.
16. Dawson, T.J. (1969), Temperature regulation and evaporative water
    loss in the brush-tailed possum *Trichosurus vulpecula*. Comp.
    Biochem. Physiol. 28 : 401-407.
17. Dawson, T.J. (1971), Thermoregulation in Australian desert kanga-
    roos. In "Climatic physiology of desert animals". Symp. Lond.
    Zool. Soc. (in press). Ed. G.M.O. Maloiy.
18. Dawson, T.J. and G.D. Brown, (1970), A comparison of the insu-
    lative and reflective properties of the fur of desert kangaroos.
    Comp. Biochem. Physiol. 37 : 23-38.
19. Dawson, T.J., M.J.S. Denny, and A.J. Hulbert, (1969), Thermal
    balance of the macropodid marsupial *Macropus eugenii* Desmarest.
    Comp. Biochem. Physiol. 31 : 645-653.
20. Dawson, T.J. and A.J. Hulbert, (1969), Standard energy metabolism
    of marsupials. Nature 221 : 383.
21. Dawson, T.J. and A.J. Hulbert, (1970), Standard metabolism, body
    temperature, and surface areas of Australian marsupials. Amer.
    J. Physiol. 218 : 1233-1238.
22. Dawson, T.J. and R.W. Rose, (1970), Influence of the respiratory
    response to moderate and severe heat on the blood gas values
    of a macropodid marsupial (*Macropus eugenii*). Comp. Biochem.

Physiol. 37 : 59-66.

23. Dawson, W.R. and J.R. Templeton, (1966), Physiological response to temperature in the alligator lizard *Gerrhonotus multicarnatus*. Ecology. 47 : 759-765.

24. Eisentraut, M. (1956), Temperaturschwankungen bei niederen Saugetieren. Zeitschr. Saugetierk. 21 : 49-52.

25. Enders, R.K. and D.E. Davis, (1936), Body temperatures of some Central American mammals. J. Mammalogy 17 : 165-166.

26. Enger, P.S. (1957), Heat regulation and metabolism in some tropical mammals and birds. Acta. Physiol. Scand. 40 : 161-166.

27. Giaja, J. (1938), X. Homeothermie et thermoregulation. II. La thermoregulation. Hermann et Cie, Paris (cited by Janský 1965).

28. Hainsworth, F.R. (1968), Evaporative water loss from rats in the heat. Amer. J. Physiol. 214 : 979-982.

29. Hammel, H.T., F.T. Caldwell, and R.M. Abrams, (1967), Regulation of body temperature in the blue tongued lizard. Science. 156: 1260-1262.

30. Hammel, H.T. (1968), Regulation of internal body temperature. Ann. Rev. Physiol. 30 : 641-710.

31. Hawkins, A.E., P.A. Jewell, and G. Tomlinson, (1960), The metabolism of some British shrews. Proc. Zool. Soc. Lond. 135 : 99-103.

32. Hayman, D.L., J.A.W. Kirsch, P.G. Martin, and P.F. Waller, (1971), Chromosomal and serological studies of the Caenolestidae and their implications for marsupial evolution. Nature 231 : 194-195.

33. Henshaw, R.E. (1970), "Thermoregulation in bats". About Bats. Slaughter and Walton, Eds. (Southern Methodist University Press, Dallas U.S.A.)

34. Higgenbotham, A.C. and W.E. Koon, (1955), Temperature regulation in the Virginia opossum. Amer. J. Physiol. 181 : 169-71.

35. Hildwein, G. (1970), Capacités thermorégulatrices d'un mammifère insectivore primitif, le Tenrec leurs variations saisonnières. Arch. Sci. Physiol. 24 : 55-71.

36. Hildwein, G. and A. Malan, (1970), Capacités thermorégulatrices du herisson en été et en hiver l'absence d'hiberbation. Arch. Sci. Physiol. 24 : 133-143.

37. Hopson, J.A. (1969), The origin and adaptive radiation of mammal-like reptiles and nontherian mammals. Ann. N.Y. Acad. Sci. 167 : 199-216.

38. Hopson, J.A. and A.W. Crompton, (1969), Origin of mammals. Evol. Biol. 3 : 15-72.

39. Irving, L., P.F. Scholander, and S.W. Grinell, (1942), Experimental studies of the respiration of sloths. J. Cell. Comp. Physiol. 20 : 189-210.

40. Ismail-Beigi, F. and I.S. Edelman, (1970), Mechanism of thyroid calorigenesis : role of active sodium transport. Proc. Nat. Acad. Sci. 67 : 1071-1078.

41. Janský, L. (1965), Adaptibility of heat production mechanisms in homeotherms. Acta. Univ. Carolinae - Biol. 1 : 1-91.

42. Johansen, K. (1961), Temperature regulation in the nine-banded armadillo (*Dasypus novemcinctus mexicanus*). Physiol. Zool. 34 : 126-144.

43. Kermack, K.A. (1967), The interrelations of early mammals. J. Linn. Soc. (Zool.) 47 : 241-249.

44. Kleiber, M. (1961), "The fire of life": An introduction to animal energetics. Wiley, New York.

45. Lasiewski, R.C. (1963), Oxygen consumption of torpid, resting, active and flying humming birds. Physiol. Zool. 36 : 122-140.

46. Lillegraven, J.A. (1969), "Late cretaceous mammals of upper part

of Edmonton formation of Alberta, Canada, and review of
marsupial-placental dichotomy in mammalian evolution".   Univ.
of Kansas Paleontological Contrib., Article 50.
47.   MacMillen, R.E. and J.E. Nelson, (1969), Bioenergetics and body
size in dasyurid marsupials.   Amer. J. Physiol. 217 : 1246-
1251.
48.   Martin, C.J. (1902), Thermal adjustments and respiratory exchange in monotremes and marsupials:- A study in the development of homeothermism.   Phil. Trans. Roy. Soc. London,
Ser. B 195 : 1-37.
49.   McKenna, M.C. (1969), The origin and early differentiation of
therian mammals.   Ann. N.Y. Acad. Sci. 167 : 217-240.
50.   McManus, J.J. (1971), Temperature regulation in the opossum
Didelphis marsupialis virginiana.   J. Mammalogy 50 : 550-558.
51.   Morrison, P.R. (1945), Acquired homiothermism in the pregnant
sloth.   J. Mammalogy 26 : 272-275.
52.   Morrison, P.R. (1946), Temperature regulation in three Central
American mammals.   J. Cell. Comp. Physiol. 27 : 125-137.
53.   Morrison, P.R. (1948), Oxygen consumption in several mammals
under basal conditions.   J. Cell. Comp. Physiol. 31 : 281-292.
54.   Morrison, P.R. (1957), Observations on body temperature in a
hedgehog.   J. Mammalogy 38 : 254-255.
55.   Morrison, P.R. (1965), Body temperatures in some Australian
mammals.   Aust. J. Zool. 13 : 173-187.
56.   Morrison, P.R. and B.K. McNab, (1962), Daily torpor in a
Brazilian murine opossum (Marmosa).   Comp. Biochem. Physiol.
6 : 57-68.
57.   Morrison, P.R. and J.H. Petajan, (1962), The development of
temperature regulation in the opossum, Didelphis marsupialis
virginiana.   Physiol. Zool. 35 : 52-65.
58.   Morrison, P.R., F.A. Ryser and A.R. Dawe, (1959), Studies on the
physiology of the masked shrew Sorex cinereus.   Physiol. Zool.
32 : 256-271.
59.   Parer, J.T. and J. Metcalfe, (1967), Respiratory studies of
monotremes.   I. Blood of the platypus (Ornithorynchus
anatinus).   Resp. Physiol. 3 : 136-142.
60.   Parrington, F.R. (1971), On the upper Triassic mammals.   Phil.
Trans. Roy. Soc. Lond. B. 261 : 231-272.
61.   Pearson, O.P. (1948), Metabolism of small mammals with remarks
on the lower limit of mammalian body size.   Science. 108 : 44-
46.
62.   Redmond, J.R. and J.N. Layne, (1958), A consideration of the
metabolic rates of some shrew tissues.   Science. 128 : 1508-
1509.
63.   Richards, S.A. (1970), The biology and comparative physiology of
thermal panting.   Biol. Rev. 45 : 223-264.
64.   Roberts, W.W., E.R. Bergquist, and T.C.L. Robinson, (1969),
Thermoregulatory grooming and sleep-like relaxation induced by
local warming of preoptic area and anterior hypothalamus in
opossum.   J. Comp. Physiol. Psych. 182-188.
65.   Robinson, K.W. and P.R. Morrison, (1957), The reaction to hot
atmosphere of various species of Australian marsupial and
placental animals.   J. Cell. Comp. Physiol. 49 : 455-478.
66.   Rudd, R.L. (1966), Body temperatures of Malaysian rainforest
mammals.   Pacif. Sci. 20 : 472-476.
67.   Rynberk, G. van, (1913), Amst. Bijdragen Dierk. 19 : 187 (cited
by Altman, P.L. and D.S. Dittmer, (1968) "Metabolism".
Federation of American Societies for experimental Biology.
(Bethesda, Maryland) p. 332.
68.   Schmidt-Nielsen, K., T.J. Dawson and E.G. Crawford, Jr., (1966),
Temperature regulation in the echidna (Tachyglossus aculeatus).

J. Cell. Physiol. 67 : 63-72.

69. Scholander, P.F. and J. Krog, (1957), Countercurrent heat exchange and vascular bundles in sloths. J. Appl. Physiol. 10 : 405-411.

70. Sharp, F., D. Smith, M. Thompson, and H.T. Hammel, (1969), Thermoregulatory salivation proportional to hypothalamic temperature above threshold in the dog. Life Sci. 8 : 1069-1076.

71. Templeton, J.R. (1970), "Reptiles". In Comparative physiology of thermoregulation. Vol. i. ed. G.C. Whittow. Academic Press, New York.

72. Wislocki, G.B. and R.K. Enders, (1935), Body temperatures of sloths, anteaters, and armadillos. J. Mammalogy. 16 : 328-329.

73. Yousef, M.K., R.R.J. Chaffee, and H.D. Johnson, (1971), Oxygen consumption of tree shrews : effects of ambient temperatures. Comp. Biochem. Physiol. 38A : 709-712.

THERMOREGULATORY BEHAVIOR

MICHEL CABANAC
Université Claude Bernard
U.E.R. Médicale Lyon-Sud-Ouest
69 - Oullins, FRANCE.

Thermoregulation is a strange function which has only recently appeared
in evolution. This function has no specific organ of its own, but it makes use of
other organs (Chatonnet, 1963). It also uses behavioral means. Behavior is a
physiological mechanism with several functions. For example, most species do not
produce enough metabolic water and therefore need a behavioral water intake ; all
animal species rely upon behavior to fulfill their metabolic substrate needs. Along
with the intake of food and water, behavior is the only means of reaching a
physiological goal.

What is thermoregulatory behavior ? It is the modification not of the
subject himself, but of his thermal environment. The subject controls his heat gain
or heat loss by changing the physical characteristics of his environment. This
includes such responses as attitude, group density, avoidance or search of thermal
environments, nesting, clothing and eventually feeding. Review articles have already
listed many kinds of thermoregulatory behavior in many species (Johansen, 1962,
Bartholomew, 1964, Hafez, 1964, Duclaux, 1970). As seen in nature, temperature
appears to be a strong behavioral determinant. All animal species have temperature-
oriented behaviors. Each species, including unicellular forms has its thermopreferendum (Viaud, 1955). If an animal with a small mass remains in an environment
with a constant temperature, its temperature will also remain constant. If the
animal can move itself in thermal gradients, a simple thermotropism with a single
thermoreceptor will therefore be sufficient to achieve the best possible thermoregulation. This seems to be the case with Paramecium aurelia (Mendelsohn, 1895)
(fig. 1). Problems increase with increasing mass. How do larger animals, with
internal heat production and long lags in heat transfer, know whether an environment

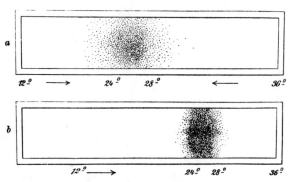

Fig. 1 : A population of Paramecium aurelia placed in an aquatic thermal
gradient moves to its thermopreferendum (above). When the preferred
temperature is displaced the animals move to stay in it (below).
(From Mendelsohn, 1895).

is favorable or not ? Do they need and/or possess one or several thermoreceptors ?
Lastly, when thermoregulation appears in homeotherms, how does this function
relate to preexisting behavior ?

Indeed, physiologists have long been unaware of how efficient behavior
can be in the thermoregulation of homeotherms. Hardy, (1961), was the first to point
out that thermoregulation is in fact mainly behavioral in terms of efficiency

(fig. 2). Actually most homeothermic species will die in spite of their
physiological defenses when placed in agressive environments.

Physiologists have focused their attention on the observation of
physiological reactions and on the nervous
system controlling these reactions. Let us
compare the nervous controls of both the
physiological and behavioral responses.

A paradox immediately appears :
apparently, the control of this behavior
relies simply upon skin sensitivity as
opposed to physiological reactions which
demand an internal control (fig. 3). Indeed,
the first physiological reaction to cold
or to warm ambient temperature is vasomotor,
increasing therefore the peripheral signal.
But with behavior, the reaction aroused
by a skin stimulus immediately corrects,
the peripheral input (fig. 4), and in itself,
this system could be regulatory if there
was a peripheral set point. Therefore, a
difference in the control of physiological
and behavioral responses appears on first
examination. This difference disappears
with a closer analysis óf the nervous
system responsible for these reactions.

HUMAN   TEMPERATURE   REGULATION

ENVIRONMENT	°C	TEMPERATURE REGULATION
VENUS	800	
	600	
	400	BEHAVIORAL REGULATION
RE-ENTERING SPACE VEHICLE		
	200	
MOON-DAY		
	80	NORMAL RANGE OF INTERNAL BODY TEMPERATURE
	60	
TROPICS	40	
	20	ZONE OF PHYSIOLOGICAL TEMPERATURE REGULATION
EARTH	0	ZONE OF THERMAL COMFORT
ARCTIC		
MOON-NIGHT MARS-SPACE	-100	BEHAVIORAL REGULATION
	-200	
ABSOLUTE ZERO	-273	

Fig.2 :Control of body temperature by physiological and behavioral
temperature regulation. (From Hardy, 1971).

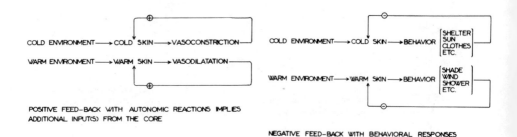

POSITIVE FEED-BACK WITH AUTONOMIC REACTIONS IMPLIES
ADDITIONAL INPUT(S) FROM THE CORE

NEGATIVE FEED-BACK WITH BEHAVIORAL RESPONSES

Fig. 3   Fig. 4

ANALOGY BETWEEN PHYSIOLOGY AND BEHAVIOR.

Physiological reactions depend upon :
- Peripheral temperature inputs : warm, cold, presumably scattered all over the body surface.
- Internal temperature inputs, hypothalamic warm and cold (Thy), infra hypothalamic CNS, other deep receptors in the abdomen, veins...
- A set level of internal temperature (Tset).
- Laws uniting the above elements, which produce a regulatory response. Various models are proposed in the literature. All of them are in agreement with Hammel's proposal (1968), that there is a proportional control of Thy. The other thermal inputs enter the system with less importance.

Let us look at behavior. Carlton and Marks (1958), then Weiss and Laties (1961) have opened a new era in developing an objective quantitative measurement of thermoregulatory behavior, by designing systems in which rats could bar press for heat (fig. 5).

a) Skin input : We have seen that it could be the sole signal. Indeed, skin stimuli give thermal sensations and in themselves, these sensations prove that thermal signals coming from the skin are capable of modifying behavior. Weiss and Laties (1961) have shown the importance of skin temperature for operant thermo-regulatory behavior in a cold environment. As we have seen, this signal in itself could be regulatory. Daily experience shows that temperature stimuli can arouse pleasure or displeasure, expressed in terms of comfort. Winslow and Herrington (1949) have observed that in man, a good correlation existed between comfort and a skin temperature of 33°C (fig. 6). Later in this chapter we shall return to the subject of skin sensation and its role in the perception of thermal comfort, after having considered how other inputs influence behavior.

b) Internal inputs : Freeman and Davis (1959) have obtained adapted attitudes of cats to the heating or cooling of the hypothalamus but Satinoff (1964) was really the first to consider the possibility of a hypothalamic role in the control of behavior. Cooling the hypothalamus of rats was followed by an adapted bar pressing for heat (fig. 7). A cold hypothalamic stimulus is therefore efficient. This result has been confirmed by many teams working on various species and a warm internal signal is also capable of triggering a corrective behavioral response (Adair, 1965 ; Baldwin and Ingram, 1967 ; Carlisle, 1966 ; Corbit, 1969 ; Gale et al, 1970 ; Hammel et al, 1967 and 1969 ; Murgatroyd and Hardy, 1970 ; Pister et al, 1967;Robinson and Hammel, 1967). The hypothalamus itself is a sensitive area and Corbit (1969) was even able to obtain hypothalamic self-stimulation in rats (fig. 8).

Experiments on infra-hypothalamic thermal stimulation and thermoregulatory behavior are still in progress. Thermal stimulation of the brain stem is not very efficient in arousing physiological reactions and evidence from thermal caudal brain stimulation is not yet clear. The posterior hypothalamus (Adair and Stitt, 1971) and the medulla (Lipton, 1971) show some capacity for controlling thermo-regulatory behavior. The spinal cord has a definite thermal sensitivity which was discovered by the Bad Nauheim team while observing physiological reactions. The first evidence that the spinal cord might also have an effect on behavior was shown in pigeons (Rautenberg, 1969, Thauer, 1970). Cooling the pigeon's spinal cord caused its legs to retract. A more systematic approach using an operant conditioned dog gave quantitative results. Cooling the spinal cord of one dog did not significantly alter its operant thermoregulatory behavior, but heating it drastically increased the response (fig. 9). The dog did not show any sign of pain with this drastic stimulus, but the specificity of this behavior is questionable since there was also a cooling behavioral response in cold environment. The spinal cord could therefore play the same role in behavioral as well as in physiological responses although the extreme temperatures necessary to obtain these reactions are somewhat puzzling. One possibility might be that all the internal thermo-detectors are efficient only as a sum of these multiple low sensitivity inputs.

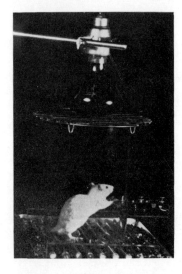

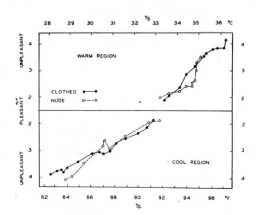

Fig. 6 : Pleasure sensations associated with skin temperature of clothed and nude subjects. (From Winslow and Herrington,1949)

Fig. 5 : The heat reinforcement apparatus designed by Weiss and Laties (1961). Depressing the lever closes a switch that activates the heat lamp.

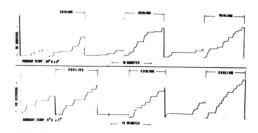

Fig. 7 : Cumulative record showing the amount of bar pressing for heat in cooling and noncooling intervals for one animal. (From Satinoff, 1964).

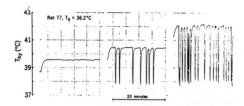

Fig. 8 : Strip-Chart records of preoptic temperature from three 20-minute periods during which skin temperature was held constant at 36.2°C. The rate of working for reductions in hypothalamic temperature (Thy) increases as hypothalamic temperature is increased. The wave form of the hypothalamic stimulus is shown.(From Corbit, 1969).

Lastly, visceral temperature should also be considered since there is a possibility that this variable might be an input with behavioral (Adair, 1971) as well as with physiological reactions (Hammel, 1968 ; Rawson and Quick, 1970 ; Bligh, 1961).

Internal temperature is also a determinant of behavior in man (Pirlet, 1962 ; Chatonnet and Cabanac, 1965 ; Benzinger, 1970). In man, the analytical techniques of exploring the nature of this internal thermal input have not reached the extensiveness permitted in animal studies. However in man, the analysis of the conscious phenomenon is possible. Taken as a whole, internal temperature plays a role by changing the way in which peripheral stimuli are perveived (Cabanac, 1969). In hypothermic subjects, warm stimuli are pleasant and cold stimuli unpleasant (fig.10). A given stimulus can therefore be pleasant or unpleasant according to the inner thermal state of the subject. The word "alliesthesia" has been proposed to describe this phenomenon. The apparent discrepancy between fig. 6 and fig. 10 is due simply to the fact that internal temperature is taken into consideration in the latter.

So far, we have seen a parallelism between physiology and behavior with regards to the inputs of the system. Does this parallelism exist when considering the thermoregulatory set point ?

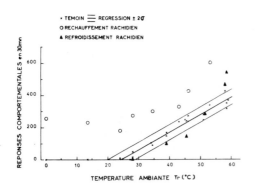

Fig. 9 : Operant behavior (ordinates) of a dog placed at various environmental temperatures (abcissae). Each point is the total number of responses given by the dog in 30 minutes. (Reinforcement = 3 sec burst of cool air) X Control with the regression curve ± 2 S.D. A U-shaped thermode is chronically implanted in the spinal canal from C3 to the sacrum. O Warm water circulated ▲ Cold water circulated. (From Cormareche and Cabanac, in preparation).

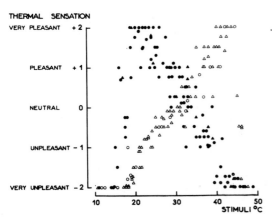

Fig. 10 : Affective responses given by a normal subject to 30 sec. thermal stimulations of the left hand. Each point corresponds to a stimulation :

△ bath at 33°C   hypothermia
○ bath at 38°C   deep body temperatures, 36.3 to 36.6°C.

▲ bath at 33°C   hyperthermia
● bath at 38°C   deep body temperatures 37.1 to 37.8°C

(From Cabanac, 1969).

COMPORTEMENT THERMOREGULATEUR

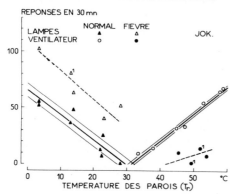

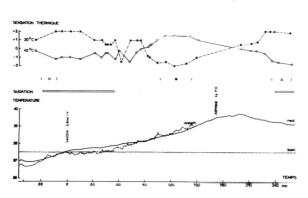

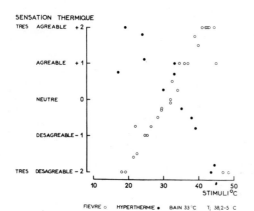

Fig. 11 : Thermoregulatory operant behavior of a dog placed at various ambient temperatures (abcissae). Each point is the total number of reinforcements obtained by the dog in the last 30 min of a 90 min experiment.
Lampes = infra red reinforcement
Ventilateur = cool reinforcement.
Regression lines $\pm$ S.D.
——— Control

----- Feverish
(From Cabanac et al, 1970).

Fig. 12 : Acute fever by intravenous injection of vaccine in a human subject. Sweating was observed on the subject's face. The two upper curves correspond to two stimuli, one warm at 40°C, the other cool at 20°C. The subject was immersed in a 37.5°C stirred water bath. Deep body temperatures were recorded in the esophagus and rectum (lower curves)
(From Cabanac, 1969).

Fig. 13 : Responses given by a feverish subject (sore throat) to thermal stimulations of the left hand. The subject was immersed in a 33°C stirred water bath. Deep body temperatures : 38.2 - 38.5°C Ti
○ Feverish
● Control
(hyperthermic)
(From Cabanac, 1969).

c) <u>Fever</u> : In a febrile state the set point is generally considered raised
to a higher level. If the set point plays any role in behavior, a feverish shift
should influence it. This is true in the case of the dog in fig. 11 which asked
for less cold reinforcements and for more warm reinforcements in the febrile states
than in the control states. Therefore, its behavior was adapted to its fever.
Experiments in man gave similar results. Acute experimental fever shifted the
preference of a slightly hyperthermic subject from cool to warm stimuli (fig. 12).
This result could be found also during a chronic pathological fever (fig. 13).
        Therefore, all elements influencing thermoregulatory physiological
response also have an effect on the control of behavior. Are these elements related
by the same laws ?

d) <u>Laws</u> : It is probably too early to answer this question. Thermoregulation models have been proposed, and so far, they are all algebraic equations,
they all use Thy, Ts and Tset - Thy and they are linear concerning Thy. Some of
them include Tset-skin. Basically, they describe physiological responses as
proportional to the difference between Tset and Thy, with a corrective factor from
the skin. They are not too different from each other if one looks at the normal
vital range ; their differences are apparent mostly at extreme temperatures and are
perhaps primarily due to the range of validity of these models.
Corbit, (1969, 1970) has proposed a model describing the thermal behavioral
responses of the rat. Stitt and al. (1971) obtained a similar model with squirrel
monkeys. Both of these behavioral models follow the above pattern for physiological
models . In man, a similar model describes the preferred hand temperature (Cabanac,
Massonnet and Belaiche, 1971). Going any further in the comparison of these models
is premature, but so far, they can be considered at least as similar.

        After reviewing thermoregulation, the two systems responsible for
physiological and behavioral thermoregulatory responses so far, appear to be
identical. It should therefore be considered that they are in fact, one system
and not two. If the physiology and the behavior are so close, they can theoreticaly
be considered as complementary. What really happens in the case of animals which
have the opportunity to choose between behavioral and physiological responses ?
Bligh (1966) proposed that thermoregulation is organised in two levels of precision
a "broad band" control then a finer control. Corbit, (1970), comparing thresholds
for physiological and behavioral responses, suggested that behavior may be a "broad
band" response and physiology a finer control. This hypothesis is supported by
Ingram and Legge (1970) who observed the thermoregulatory behavior of young pigs
in a natural environment and concluded that these animals "tolerated some degree
of thermal discomfort before modifying their behavior especially when the
discomfort is endured for the sake of obtaining food". The same conclusion can be
drawn from the results of Stitt and al (1971) on the squirrel monkey.
        This theory should perhaps be questionned, at least in dogs. Some dogs
showed a very well-adapted behavioral response, little spreading of experimental
points and reproducible results (fig. 14). Their behavior was accompanied by
physiological responses : shivering in a cold environment and panting in a warm
environment. They followed the pattern of behavioral broad band and physiological
fine controls. In contrast, a second group of dogs, in exactly the same conditions
as the other group, showed a different behavior pattern : i) their responses were
variable with a spread of experimental points, ii) cold and warm responses
overlapped each other (fig. 15). However, they did not shiver nor pant. They used
their behavior to spare their physiological responses. Apparently, they showed the
opposite pattern with a fine behavioral control. This trend seems to be frequent
among birds (Bartholomew, 1964). Also, man will usually avoid sweating and
shivering if he is able to modify his clothing and/or his environment. It is
therefore reasonable to admit that physiological and behavioral responses are used
equally depending only on circumstances. The superiority of the homeotherm over the
poïkilotherm is precisely this free choice between behavior and physiology in
moderate environments. The choice of one response allows the other response to
be free for other purposes. This choice is probably innate and not learned since
it exists already at birth (Mount, 1963).

DIFFERENCE BETWEEN PHYSIOLOGY AND BEHAVIOR :

This equality and complementarity of physiology and behavior will appear in the
observations made after lesions were inflicted on the CNS. These experiments will
pose a very intriguing problem if one admits the identity of the control of
physiological and behavioral thermoregulatory responses.

        a) Lipton (1968) has observed that rats with rostral lesions of their
hypothalamus cannot maintain their body temperature at high ambient temperatures.
This is a new observation of the well known impairment of temperature regulation
with this kind of lesion. Remarkably, when offered the possibility of using an
instrumental heat escape bar pressing system, these animals used it more than in
the control periods (fig. 16). This is particularly strange in a species which
relies essentially on saliva spreading, a type of a behavior, as its normal
response. The same phenomenon in a cold environment was observed in its crude from
by Rudiger and Seyer (1965) then by Carlisle (1969) and by Satinoff and Rutstein
(1970). Physiological responses were impaired in rats with bilateral hypothalamic
lesions. They had a tendency towards hypothermia but, at the same time, operated
rats increased their behavioral response rate for radiant heat (fig. 17). Therefore
in these experiments, behavior tended to compensate the physiologic deficit in
both cold and warm environments. If this situation is satisfactory from a
finalistic point of view, it presents a puzzling problem on how the internal need
is perceived in the absence of hypothalamic sensors. These results imply that
internal temperature is detected from other sources than the hypothalamus. We
have seen that they do exist and it is possible that these extra hypothalamic
thermal detectors are at the origin of this behavior. However, these results also
imply that not all the integration is done in the preoptic area at least concerning
behavior. The nervous network responsible for obtaining behavioral reactions is
therefore different from the system responsible for physiological reactions. The
thermoregulatory function uses all means at its disposal in order to reach its
goal. It should be pointed out that a similar situation exists with regard to
water intake where the response is exclusively behavioral. There is probably a
primary polydipsia when a subject develops diabetes insipidus following a hypothalamic lesion. However, there is no doubt that a secondary polydipsia also
corrects the inability of his kidneys to concentrate urine, although his osmo-
receptors had been damaged at the same time as his ADH secretory cells.

        b) The opposite situation was found in two patients who were indifferent
to pain (fig. 18) (Cabanac et al, 1989). In these subjects, the alliesthesic shift
of thermal sensations for a whole range of peripheral stimuli did not show up,
although their internal temperatures were changed by the bath method. Thus, these
two subjects, without alliesthesia, did not report nor show any sign of thermal
discomfort during the whole course of the experiments, although their core
temperatures were carried from 36.15°C to 38.5°C. These two subjects did not show
any trouble in their physiological thermoregulation. They maintained 37°C in their
normal life, shivered in the cool bath and sweated when hyperthermic in the warm
bath. This confirms the duality of the systems responsible for physiological
responses, and thermal motivation.

Another difference between physiology and behavior exists in poïkilotherms and some
newborn homeotherms. In these two cases, physiological reactions do not exist or
are not yet mature, however, thermoregulation exists and expresses itself through
behavior. Many examples of thermoregulatory behavior in poïkilotherms have been
cited, but as pointed out above, it is difficult to separate thermotropism from
thermoregulation in animals with a small mass. Nevertheless, in larger poïkil
lotherms thermal behavior is regulatory. There is a great similarity in the
network responsible for thermoregulation in poïkilotherms and in homeotherms
(Hammel et al 1967-1969). With newborns, behavior exists (Bruck et al., 1962) before
the maturation of physiological heat loss control, as is seen by efficient
huddling (Cosnier et al, 1965) or by the search of the mother's fur (Jeddi, 1970).

COMFORT :

Each element of thermoregulation has revealed a behavioral counterpart, skin

COMPORTEMENT THERMOREGULATEUR

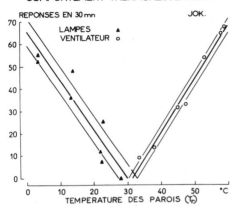

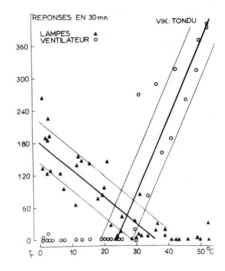

Fig. 14 : Thermoregulatory operant behavior of one dog. Each point is the total number of responses given by the dog at different environmental temperatures during the last 30 minutes of a 90-minute experiment.

▲ Infra red reinforcement
o Cool air reinforcement
— Regression lines ± S.D.

(From Cabanac et al., 1970).

Fig. 15 : Same legend as fig. 14.

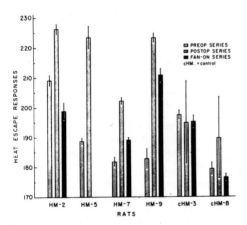

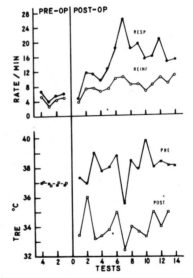

Fig. 16 : Mean (±S.E.) number of heat-escape responses per one-hr session, made by experimental and control rats during the preoperative (15 days) post-operative (15 days) and fan-on (15 days) series. (From Lipton, 1968).

Fig. 17 : Mean rates of response and reinforcement pre- and postoperatively (hypothalamic ablation), and pre- and post - test rectal temperature (Tre) for rat 72. (Reinforcement duration = 2 sec).

(From Carlisle, 1969).

input, internal inputs, set point and laws. Let us try to analyse the phenomenon of consciousness commonly described as thermal comfort that allows man and presumably animals to know whether a given environment is beneficial or dangerous. It is difficult to speak of the thermal comfort of a <u>Paramecium</u> but we may assume that bigger animals have a perception of thermal comfort. Is comfort the origin of thermoregulatory behavior ? To analyze this feeling let us look at man. By personal experience everybody knows what thermal comfort or discomfort is, but the analysis of the origin of this feeling is not as evident (see Chatonnet and Cabanac, 1965, Hardy, 1970). Perhaps in heat discomfort, vascular changes should be considered. Thermoregulatory physiological responses such as wet skin, panting, and shivering are conscious and might also participate in the perception of discomfort. But as Corbit (1970) points out, in poïkilothermic species, these reactions do not exist, but they still have a thermoregulatory behavior. The same is true in new born homeotherms. Spinal cord cooling created a violent shivering and the animal did not try to obtain radiant heat. Therefore, even if physiological reactions participate in thermal discomfort, it is probable that they have a minor role in this feeling compared to the importance of pleasure in thermal sensation. Marks, (1971) ; Stevens and Banks, (1971), have described the laws governing the summation of skin stimuli to produce thermal sensation. Since pleasure appears with a thermal stimulus when it is useful with regard to the inner thermal state, it is likely that the summation of this effect to the total body surface is thermal comfort or discomfort. I would therefore offer to slightly change Gagge's (1971) and Ashrae's definition of thermal comfort from " the condition of mind which expresses satisfaction with the thermal environment", to the slightly more conservative definition, "the absence of unpleasant feeling" because satisfaction can be either the absence of displeasure or the presence of pleasure. This hypothesis of alliesthesia-dependent comfort, would explain the comfort experienced by subjects stimulated both with cold and heat on large skin surfaces (Bøje et al, 1948 ; Kaletzky et al, 1963 ; Hall and Klemm, 1969). The problem of the direct perception of the inner temperature can perhaps be explained by this effect. Corbit (1969) has obtained intracranial thermal self stimulation in rats, but hypothalamic temperature is not necessarily perceived by these rats. Perhaps, this signal simply makes the peripheral sensations seem pleasant to these animals. Additional experiments provide other reasons to believe that internal temperature per se is not perceived and has to go through alliesthesic changes in peripheral sensation in order to become conscious. Murgatroyd and Hardy (1970) measured the bar pressing rates for cold or for heat in rats at various ambient temperatures while simultaneously stimulating the hypothalamus with cold or warm thermodes (fig. 19). It is striking that their results show no effect from hypothalamic stimulations at neutral temperatures. It is, as if the internal signal was inefficient in the absence of a peripheral thermal signal. The case of the two patients reported above also induces one to think that no thermal discomfort can be perceived in the absence of alliesthesia. This observation again gives the main if not the only role to pleasure and displeasure in the perception of thermal comfort. It is more tempting to think that the search for pleasure is the main motivation for thermoregulatory behavior, because this factor is very efficient. Fig. 20. shows an example of a very pleasant sensation produced by immersing the hand of a hyperthermic subject in cool water. This pleasant stimulus extracts from the hand as much as 80 W which is approximately the heat production of the B.M.R. It is interesting to note that the hyperthermic beaver has a behavior quite similar to the above subject since it dips its tail in cool water to prevent further hyperthermia (Steen and Steen, 1965).

BEHAVIOR REACTIONS USED AS PHYSIOLOGICAL INDICATORS :

        Finally, in spite of the discrepancies seen in pathology or after hypothalamic lesions, behavior or comfort are closely related to physiological responses in normal circumstances and on the intact animal and both have the same goal. This relation is so close that behavioral responses can be used as an index of the difference between actual inner temperature and set temperature. Therefore, as with physiological responses, behavioral responses will allow the exploration of

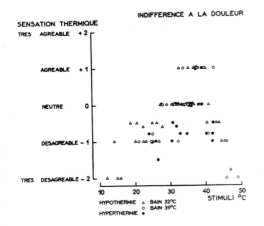

Fig. 18 : Affective responses (ordinates) to thermal stimulations (abcissae) in a patient indifferent to pain. Compare this figure with fig. 10, which is the same experiment on a normal subject.

hypothermia △ Bath 32°C
              O Bath 39°C
hyperthermia ●

(From Cabanac et al, 1969).

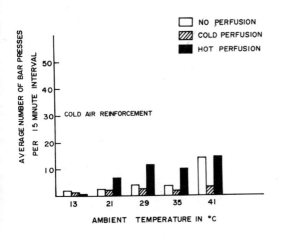

Fig. 19 a : Average bar-press rate for cold-air reinforcement during thermal clamping of the preoptic area at ambients 13 to 41°C. Reinforcement 8 sec cooling.

(From Murgatroyd and Hardy, 1970).

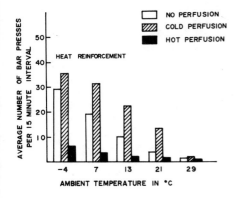

Fig. 19 b : Average bar-press rate for heat reinforcement during thermal clamping of the preoptic area at ambients -4 to 29°C. Reinforcement 2 sec heating.

(From Murgatroyd and Hardy, 1970).

the set point for temperature regulation in normal intact animals or subjects.This
assumption implies that behavior may be as fine a control as physiology and not
necessarily limited to a broad band control. Behavior has been used as an index
of the set temperature in three circumstances : after intra cerebral injections
of noradrenaline and acetylcholine, during muscular exercise and during the
menstrual cycle.

Intra cerebral injections of noradrenaline at various ambient temperatures were
followed by increases in hypothalamic temperature in rats (Beckmann, 1970) and
by either increases or decreases in the hypothalamic temperature in dogs (Duclaux,
Cabanac, 1971). The thermoregulatory behaviors of these animals were also modified.
All temperature drops were accompanied by a cooling behavior. Intra-cerebral
acetylcholine was followed by a decreased hypothalamic temperature and a reduced
heating behavior (Beckmann and Carlisle, 1969). These results confirm that
behavior is closely adapted to any thermoregulatory change, although none of the
current theories about amines and set-point can be confirmed since internal
temperature shifts following intra-ventricular nor-adrenaline injections were
either raises or drops and were independent from ambient temperatures.
        More information could be gained from the observation of women during
the menstrual cycle (Cunningham and Cabanac, 1971). During the preovulatory
phase of the cycle, morning rectal temperature is 0.4 to 0.5°C below post
ovulatory rectal temperature measured in the same conditions. One can question
the nature of this shifting temperature. Is it passive, due to an excess or a
deficit in heat production or in heat conservation during one phase or the other,
or is it a regulated phenomenon ? If one agrees that pleasure or displeasure given
by a thermal skin stimulus depends upon the difference between set temperature
and real internal temperature, the measurement of internal temperature and of
thermal pleasure in response to given stimuli, will indicate the set temperature.
Results showed that the difference in rectal temperature between pre-and post-
ovulatory periods corresponds to different settings of the biological thermostat
(fig. 21).

        Another circumstance where behavior was used as a means of exploring
the thermostat was muscular exercise, (Cabanac, Cunningham, Stolwijk, 1971).
During exercise, the internal temperature is elevated.At present they are several
opposing theories based on apparently contradictory facts. The problem is to
know whether non-thermal inputs reset the biological thermostat to a higher level,
to a lower level or do not change it at all. A method quite close to the
proceeding example permitted a contribution to the problem. Subjects were equipped
with a rubber glove in which water was circulated at the temperature chosen by
the subjects themselves. This preference was used as a sign of hypo- or
hyperthermia depending on the choice of warm or of cold water. This method was
used to observe the choice during a short period of muscular exercise on an
ergometric bicycle (fig. 22). The results showed no change in hand temperature
preference neither at the start of exercise nor at the end, suggesting that no
non-thermal input enters the system to reset the thermostat. The only changes
in glove temperature observed could be explained by the heat production in
muscular work and the subsequent elevation of internal temperature, followed
by the return to control conditions.

CONCLUSIONS :

        Thermoregulation is accomplished through behavior as well as through
physiological reactions. Both of these reactions seem to be complementary.
However, the nervous networks responsible for motivation and physiological
reactions are different. In intact animals and humans, the signals for thermoregulatory behavior are thermal discomfort and thermal comfort.

        The final remarks of this chapter will be concerned with semantics. The
currant literature communly speaks of "behavioral thermoregulation" as
implicitly opposed to "physiological thermoregulation". As we have seen,
physiology and behavior are closely related. If, in some special circumstances they

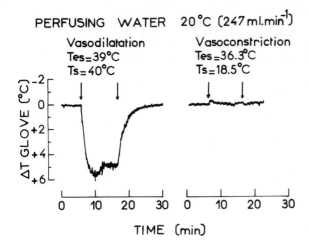

PERFUSING WATER   20°C (247ml.min^{-1})

Vasodilatation
Tes=39°C
Ts=40°C

Vasoconstriction
Tes=36.3°C
Ts=18.5°C

ΔT GLOVE (°C)

TIME (min)

Fig. 20 : Example of the thermal efficiency of a very pleasant sensation : the left hand of the subject was dipped in a glove perfused with water. Δ T is the input-output temperature difference.

(From Cabanac, Massonnet, Belaiche, unpublished).

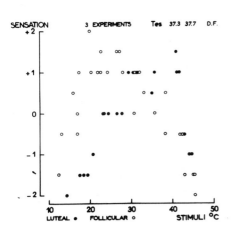

SENSATION

3 EXPERIMENTS   Tes 37.3 37.7   D.F.

LUTEAL ●   FOLLICULAR ○   STIMULI °C

Fig. 21 : Affective responses to thermal stimuli given by one female subject in three experiments conducted during follicular and luteal phases of the cycle. The subject had the same skin temperature (37.5°C) and the same internal temperature (Ts = 37.2 - 37.4°C).

(From Cunningham and Cabanac, 1971).

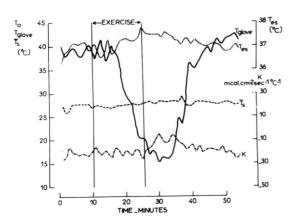

Ta
Tglove
Ts
(°C)

Tes (°C)

mcal.cm^{-2}sec^{-1}°C^{-1}

TIME _ MINUTES

Fig. 22 : Temperature preferred by a subject (T glove) placed at cool ambient temperature, before, during and after a short period of ergometric bicycle exercise.
The temperature preference does not seem to be influenced by the onset nor by the end of exercise, but rather by internal temperature.

(From Cabanac, Cunningham, Stolwijk, 1971).

can be dissociated, they are still complementary in normal life and it therefore
seems arbitrary to consider them opposites. Epstein, then Bligh[x] have pointed
out that both are physiological for the maintainance of internal temperature.
Several solutions to this semantic problem have been proposed to eliminate this
opposition. Epstein [x], then Duclaux (1970) proposed "behavioral" versus "automatic"
and Benzinger (1970), and Kayser[x], proposed "behavioral" versus "autonomic".
Neither prpposition is exact and perhaps neither will be accepted into common use.
The problem becomes less serious if physiology and behavior are used as
substantives instead of adjectives. Instead of speaking of physiological
thermoregulation and behavioral thermoregulation, one should speak of thermoregulatory behavior and thermoregulatory physiological responses. This practice has
been adopted in this chapter.

x Personal communication.

Adair, E.R., (1971),Displacements of rectal temperature modify behavioral
          thermoregulation, Physiol. Behav. 7 (1), 21.

Adair, E.R., J.U. Casby and J.A.J. Stolwijk, (1970), Behavioral temperature
          regulation in the squirrel monkey : changes induced by shifts in
          hypothalamic temperature, J. Comp. Physiol. Psychol. 72 (1), 17.

Adair, E.R. and J.T. Stitt, (1971), Behavioral temperature regulation in the
          squirrel monkey : effects of midbrain temperature displacements,
          J. Physiol. (Paris), 63 (3), 191.

Baldwin, B.A. and D.L. Ingram, (1967), The effect of heating and cooling the
          hypothalamus on behavioral,thermoregulation in the pig, J. Physiol.
          (London), 191, 375.

Bartholomew, G.A., (1964), the roles of physiology and behaviour in the maintenance
          of homeostasis in the desert environment, (Cambridge University Press)
          p. 7.

Beckmann, A.L., (1970), Effect of intrahypothalamic norepinephrine on thermoregula-
          tory responses in the rat, Am. J. Physiol., 218 (6), 1596.

Beckman, A.L. and H.J. Carlisle, (1969), Intrahypothalamic infusion of acetyl-
          choline in the rat, Nature, 221, 561.

Benzinger, T.H., (1970), Peripheral cold reception and central warm reception
          sensory mechanisms of behavioral and autonomic thermostasie, Physiol.
          and Behav. Temperature Regulation, J.D. Hardy, A.P. Gagge, J.A.J.
          Stolwijk, chap. 56, 831.

Bligh, J., (1961), Possible temperature sensitive elements in or near the vena
          cava of sheep, J. Physiol. (London), 159, 85.

Bligh, J., (1966), the thermosensitivity of the hypothalamus and thermoregulation
          in mammals, Biol. Rev., 41, 317.

Bøje, O., M. Nielsen and J. Olesen, (1948), Undersøgelser over hetydningen of
          ensidig straalingsafkøling, Boligopvarmningsudvalgets Med. n°9
          (Copenhague).

Bruck, K., A.H. Parmelee and M. Bruck, (1962), Neutral temperature range and range
          of "thermal comfort in premature infants", Biol. Neonate, 4 (1-2), 32.

Cabanac, M., (1969), Plaisir ou déplaisir de la sensation thermique et homéothermie,
          Physiol. Behav., 4, 359.

Cabanac, M., P. Ramel, R. Duclaux and M. Joli, (1969), Indifférence à la douleur et
          confort thermique. Etude expérimentale de deux cas, Presse Médicale
          (Paris), 77.(54), 2053.

Cabanac, M., R. Duclaux and A. Gillet, (1970), Thermorégulation comportementale
          chez le chien. Effet de la fièvre et de la thyroxine, Physiol. Behav.,
          5 (6), 697.

Cabanac, M., D.J. Cunningham and J.A.J. Stolwijk, (1971), Thermoregulatory set
          point during exercise : a behavioral approach, J. Comp. Physiol. Psychol.
          75 (3),

Cabanac, M., B. Massonnet and R. Belaiche, (1971), Preferred hand temperature as a
          function of internal and mean skin temperatures, (Abstracts Internat.

34   MICHEL CABANAC

Sympos. Temp. Regul. - Dublin), p. 51.

Carlisle, H.T., (1966), Behavioural significance of hypothalamic temperature-sensitive cells. Nature, 209, 1324.

Carlisle, H.T., (1969), Effect of preoptic and anterior hypothalamic lesions on behavioral thermoregulation in the cold, J. Comp. Physiol. Psychol., 69 (2), 391.

Carlton, P.L. and R.A. Marks, (1958), Cold exposure and heat reinforced operant behavior, Science, 69, 1344.

Chatonnet, J., (1963), Régulation thermique in Physiologie (Kayser), chap. 13, 1264.

Chatonnet, J. and M. Cabanac, (1965), The perception of thermal comfort, Internat. J. Biomet., 9 (2), 183.

Corbit, J.D., (1969), Behavioral regulation of hypothalamic temperature, Science, 166, 256.

Corbit, J.D., (1970), Behavioral regulation of body temperature, Physiol. and Behav. Temperature Regulation, J.D. Hardy, A.P. Gagge, J.A.J. Stolwijk, chap. 53, 777.

Cosnier, J., A. Duveau and J. Chanel, (1965), Consommation d'oxygène et grégarisme chez le rat nouveau-né, C.R. Soc. Biol., 159, 1579.

Cunningham, D.J. and M. Cabanac, (1971), Evidence from behavioral thermoregulatory responses of a shift in set point temperature related to the menstrual cycle, J. Physiol. (Paris), 63 (3), 236.

Duclaux, R., (1970), Thermoregulation comportementale chez le chien. Etude expérimentale., Thèse Doctorat Biol. hum. Univ. Lyon, p. 207.

Duclaux, R. and M. Cabanac, (1971), Effets de la noradrénaline intracérébrale sur le comportement thermorégulateur chez le chien, J. Physiol. (Paris), 63 (3), 246.

Freeman, W.J. and D.D. Davis, (1959), Effets du refroidissement hypothalamique par conduction sur des chats, Am. J. Physiol., 197, (1), 145.

Gagge, A.P., (1971), Research on thermal comfort; A physiologist 's view, Ashrae J.

Gale, C.C., M. Mathews, and J. Young, (1970), Behavioral thermoregulatory responses to hypothalamic cooling and warming in baboons. Physiol. Behav., 5, 1.

Hafez, E.S.E., (1964), Behavioral thermoregulation in mammals and birds (a review) Inter. J. Biomet., 7 (3), 231.

Hall, J.F. and F.K. Klemm, (1969), Thermal comfort in disparate thermal environments J. of Applied Physiol., 27 (5), 601.

Hammel, H.T., (1968), Regulation of internal body temperature, Ann. Rev. Physiol., 30, 641.

Hammel, H.T., F.T. Caldwell Jr., and R.M. Abrams, (1967), Regulation of body temperature in the blue-tongued lizard, Science, 156, 1260.

Hammel, H.T., S.B. Strømme, and K. Myhre, (1969), Forebrain Temperature Activates behavioral thermoregulatory response in arctic sculpins, Science, 165,83.

Hardy, J.D., (1961), Physiology of temperature regulation, Physiol. Rev., 41 (3), 521.

Hardy, J.D., (1970), Thermal comfort : skin temperature and physiological thermo-regulation, Physiol. and Behav. temperature regulation, J.D. Hardy, A.P. Gagge, J.A.J. Stolwijk, chap. 57, 856.

Hardy, J.D., (1971), Thermal comfort and health, Ashrae J., p. 43.

Ingram, D.L. and K.F. Legge, (1970), the thermoregulatory behavior of young pigs in a natural environment, Physiol. Behav., 5, 981.

Jeddi, E., (1970), Confort du contact et thermorégulation comportementale, Physiol. Behav., 5, 1487.

Johansen, K., (1962), Evolution of temperature regulation in mammals, Comp. Physiol. of temperature regulation, (J.P. Hannon - E. Viereck. ed. AAL Lab. Fort W ainwright), 73-131.

Kaletzky, E., R.K. Macpherson and R.N. Morse, (1963), the effect of low temperature radiant cooling on thermal comfort in a hot, moist environment, Electr. Mech. Eng. Trens. (Austral), 5 (2), 60.

Lipton, J.M., (1968), Effect of preoptic lesions on heat-escape responding and colonic temperature in the rat., Physiol. and Behav., 3, 165.

Lipton, J.M., (1971), Thermal stimulation of the medulla alters behavioral temperature regulation, Brain Research, 26 (2), 433.

Marks, L.E., (1971), Spatial summation in relation to the dynamics of warmth sensation, (Abstracts Internat. Symp. Temp. Regul. Dublin) p. 69.

Mendelssohn, M. (1895), Ueber den thermotropismus einzelliger Organismen, Arch. ges. Physiol., 60, 1.

Mount, L.E., (1963), Environmental temperature preferred by the young pig, Nature, 199, 1212.

Murgatroyd, D. and J.D. Hardy, (1970), Central and peripheral temperatures in behavioral thermoregulation of the rat, Physiol. and Behav. temperature Regulation, J.D. Hardy, A.P. Gagge, J.A.J. Stolwijk, chap. 58, 874.

Pirlet, K., (1962), Die Verstellung des Kerntemperatur. Sollasestes bei Kältebelastung, Pflügers Archiv, 275, 71.

Pister, J.D., M. Jobin and C.C. Gale, (1967), Behavioral responses to anterior hypothalamic cooling in unanaesthetized baboon. Physiologist, 10, 279.

Rautenberg, W., (1969), Die Bedeutung der zentralnervösen Thermosensitivitat für die Temperatur regulation der Taube, Zeischrift für vergleichende Physiologie, 62, 235.

Rawson, R.O. and K.P. Quick, (1970), Evidence of deep body thermoreceptor response to intra-abdominal heating of the ewe, J. of Applied Physiol., 28, (6), 813.

Robinson, J.J. and H.T. Hammel, (1967), Behavioral thermoregulation in response to heating and cooling of the hypothalamic preoptic area of the dog. Rept. AMRL-TR-67-144. Wright Patterson, AFB Ohio.

Rudiger, W. and G. Seyer, (1965), Lateralisation of cortico-hypothalamic relations
          as revealed by thermosensitive behaviour in the rat, Physiol. Biochem.,
          14, 515.

Satinoff, E., (1964), Behavioral thermoregulation in response to local cooling of
          the rat brain. Am. J. Physiol., 206 (6), 1389.

Satinoff, E. and J. Rutstein, (1970), Behavioral thermoregulation in rats with
          anterior hypothalamic lesions, J. of Comp. and Physiol. Psychol., 71 (1)
          77.

Steen, I. and J.B. Steen, (1965), Thermoregulatory importance of the beaver 's tail
          Comp. Biochem. Physiol., 15 (2), 267.

Stevens, J.C. and W.P. Banks, (1971), Spatial summation in relation to speed of
          reaction to radiant stimulation. (Abstracts Intern. Symp. Temp. Regul.
          Dublin) p. 88.

Stitt, J.T., E.R. Adair, E.R. Nadel and J.A.J. Stolwijk, (1971), The relation
          between behavior and physiology in the thermoregulatory response of the
          squirrel monkey, J. Physiol. (Paris), 63 (3), 424.

Thauer, R., (1970), Thermosensitivity of the spinal chord, Physiol. and Behav.
          temperature regulation, J.D.Hardy, A.P. Gagge, J.A.J. Stolwijk, chap.33,
          472.

Viaud, G., (1955), Le thermopréférendum et les thermotropismes, Coll. Nat. C.N.R.S.
          Strasbourg, p. C 35.

Weiss, B. and V.G. Laties, (1961), Behavioral thermoregulation, Science, 133, 1338.

Winslow, C.E.A. and L.P. Herrington, (1949), Temperature and Human life, Princeton
          University Press, p. 104.

MATHEMATICAL AND PHYSICAL MODELS
OF THERMOREGULATION

DUNCAN MITCHELL*
A.R. ATKINS**
C.H. WYNDHAM*

*Human Sciences Laboratory,
**Physical Sciences Laboratory,
 Chamber of Mines of South Africa,
 Johannesburg, South Africa.

The usefulness of modelling can be demonstrated by means of a simple model of thermoregulation. The model shown in Figure 1 is naive, but it demonstrates some of the salient features of thermo-regulation. The figure shows a water cistern with ball valves controlling both an inlet to and an outlet from the cistern. The floats sense and control the level of the water. The cistern is subjected to two kinds of disturbance: emptying and over-filling. During emptying the float controlling the inlet valve detects a drop in water level. The information is fed back mechanically to the control valve which opens to allow more water into the tank. Simil-arly if water leaks into the cistern the outlet control valve will open and allow the excess water to escape.

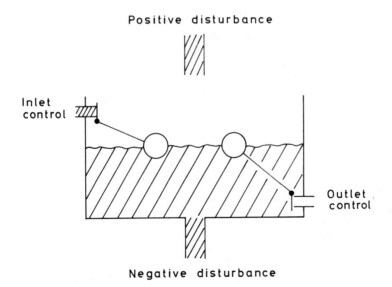

Fig. 1.   Two-valved water cistern – a simple model of
          thermoregulation.

Some of the ways in which the operation of the cistern represents thermoregulation may be elaborated. For instance, when the cistern is subjected to a continuous positive disturbance, that is, a continuous leak into the cistern, the water will reach a new steady level when there is a dynamic equilibrium between the rate of inflow and the rate of outflow. For the maintenance of the new equilibrium the outlet control must be open and therefore the float controlling this valve must be elevated above its previous steady level. By comparison, in thermoregulation, a continuous heat load on the body, such as metabolic heat production, must be associated with an elevated but steady body temperature. The simple model also demonstrates how changes in setpoint and changes in gain or sensitivity both result in changes in the operating level of the controlled variable. A change of setpoint corresponds to a change in the vertical position of the control valves, clearly causing a change in water level. A change of gain corresponds to a change in the relative lengths of the arms between the floats and the valves, which also results in a change in water level.

The cistern model is a typical qualitative model of a complex biological process. In the words of MacDonald and Wyndham (1950) it "crystallizes in one diagram the characteristics of a system which otherwise requires much apparently complex mathematical or verbal explanation". It is not an analogue. Two systems are analogous only if there is a correspondence between each dependent variable and its derivatives in the two systems (Atkins, 1962). For two systems to be analogous they must be described by the same mathematical equations. Only then can quantitative conclusions be made about one of the systems by studying the other system (Dainty, 1960). The cistern model is a physical (or engineering) model rather than a mathematical model because it compares the biological system with a more familiar physical system.

MATHEMATICAL MODELS

Kac (1969) defines a mathematical model to be a model "which can be described symbolically and discussed deductively". Mathematical models describe the behaviour of a system, rather than describing the system.

Pure mathematical models of thermoregulation do exist. Perhaps the first and still the most widely used is the symbolic expression of the First Law of Thermodynamics:

$$M + W + R + C + E = S$$
$$= 0 \text{ in the steady state}$$

(M: rate of metabolic energy consumption; W: work rate; R, C, E,: rates of radiant, convective and evaporative heat transfer with the environment respectively; S: rate of heat accumulation; sign convention: energy transfer tending to raise body temperature is positive). The equation states that when a body is subjected to a thermal disturbance, heat is stored or lost. After some time a new equilibrium will be set up when the net rate of storage is zero. The symbolic equation is a fundamental descriptive mathematical model (Kac, 1969) of thermoregulation, based on a well-known physical law.

The equation implies that if during thermal equilibrium M, W, R and C are known, E is also known. To say that sweat rate in man can be predicted with high accuracy from metabolic rate and ambient temperature (Stolwijk, 1968b) reveals nothing, however, of how sweating is controlled in man. The model specifies a necessary quantitative relationship between the various energy flows in and out of a "black box" (Hardy, 1965). The relationship must hold true

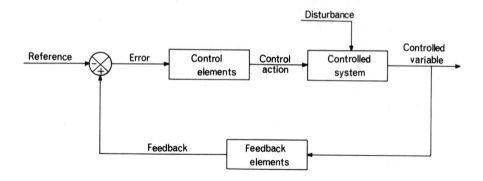

Fig. 2.   Basic negative feedback control system with a reference
          signal. (From D. Mitchell, J.W. Snellen, and A.R. Atkins,
          (1970), Arch. Ges. Physiol. <u>321</u>, 293).

irrespective of how the components of the black box function.
   The block diagram is another form of mathematical model, a form
with a clarity not found in algebraic mathematical models. Block
diagrams provide a good basis for organizing the analysis of a
system.   They describe the significant variables, the interrelationship of these variables and the dynamic characteristics of
the system elements.
   Block diagrams can be constructed for any form of control system.
Thermoregulation generally conforms to the concepts of negative
feedback control and Figure 2 is a block diagram of a basic negative
feedback control system.   Each block represents the behaviour of a
particular portion of the entire system.   The mathematical
expression relating the output to the input of each block is termed
its "transfer function", or, more properly, its "coupling function"
(Hardy, 1968).   The mathematical expression describing the
behaviour of the whole system is obtained by combining the transfer
functions of all the blocks in the sequence in which they are
connected.
   Let us examine how thermoregulation conforms to the block diagram
of Figure 2.   The diagram contains a controlled system which in the
case of autonomic thermoregulation is the body of the homeotherm.
The controlled system can be subjected to a disturbance: environmental heat or cold, or metabolic heat.   The disturbance causes a
change in a particular controlled parameter, the controlled variable,
presumably a body temperature or combination of temperatures.   The
controlled variable is measured by a transducer which generates
related neural or humoral information.   This information, the feedback, is compared with some reference information.   The difference
between the feedback and the reference, termed the error, is a
measure of the effect of the disturbance on the controlled variable.
The difference activates a control centre which provides a control
action in such a way as to oppose the effect of the disturbance.
In thermoregulation the control actions are means of accelerating
heat loss, heat production or heat conservation (such as sweating or
shivering).
   Block diagrams like that of Figure 2 cope only with short-term
exposures to heat, cold or exercise.   Diagrams coping with longer
term adaptive changes require at least one extra loop.   Such a
diagram has been constructed by Stolwijk and Hardy (1966).   Block
diagrams which attempt to describe in more detail mechanisms simply

40   DUNCAN MITCHELL, A. R. ATKINS, C. H. WYNDHAM

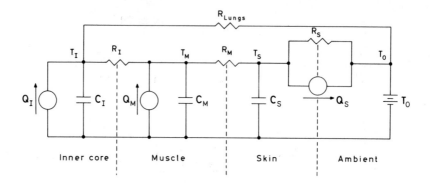

Fig. 3.   Mac Donald-Wyndham model in improved form.   (Redrawn
         from R.W. Cornew, J.C. Houk, and L. Stark, (1967),
         J. Theoret. Biol. **16**, 406).

enclosed in black boxes in diagrams like Figure 2 become very complex
(Hammel, 1968).

Simplicity in block diagrams, and models in general, is not
necessarily a disadvantage (Dainty, 1960).   One can consider a
simple block diagram to be a building block of a much larger model
which, in the end, must take account not only of all aspects of the
thermoregulatory control system but also its interactions with cardiovascular, respiratory, endocrine and body fluid control systems
(Stolwijk, 1966; Brown, 1970).   One has to understand each building
block before it can be combined with others.

PHYSICAL MODELS

Though mathematical models have the advantage of being quantitative they are by definition abstract, and therefore do not fulfil
one of the functions of models, that of presenting complex phenomena
in terms of familiar concepts.   Physical models do fulfil this
function.   The first physical model of thermoregulation was probably
the electrical model of MacDonald and Wyndham published in 1950.
MacDonald and Wyndham (MacDonald, 1950; Wyndham 1952) were trying to
demonstrate the role of active control and the interrelationship of
body masses, internal resistances and external resistances in the
response of men to work and heat.   They devised an electrical model
which is shown in the improved form due to Cornew et al. (1967) in
Figure 3.

MacDonald and Wyndham employed the analogy between rate of heat
flow and electrical current flow and assigned the well-known
electrical symbols to thermal capacitance, thermal resistance and
thermal potential (temperature) difference.   Masses, and therefore
capacitances, were accorded to various body regions.   Heat flowing
between regions encountered thermal resistance.   Metabolically
active regions were modelled by current sources.   The environment
was represented by a potential difference.   Heat travelling between
the body and the environment met resistance representing the
resistance to radiation and convection.   A current source represented evaporative heat transfer.   The various resistances in the model
were variable and dependent on body temperature.

The electrical model assists those familiar with simple electrical
circuits to understand how the temperature in a particular region of

the body could be expected to behave during metabolic and environmental heat loads.   Because the resistances R_I and R_M are dependent on blood flow and distribution, the influence of the state of the circulatory system on thermal reactions can be predicted. Although the model was only semi-quantitative, MacDonald and Wyndham estimated thermal time constants for the human body by alloting values to the various resistances and capacitances.

In his Harvey lecture of 1954, Hardy (1953/4) described an electrical model similar to the MacDonald-Wyndham model and also mentioned that other physical models involving "tanks of water, thermostats, evaporating surfaces, and curculating water systems" had been used for many years.   None of these other physical models seems to have survived.   In comparison with electrical models, mechanical, thermal and hydraulic models are bulky, difficult to build, costly, inaccurate and lacking in versatility (Atkins, 1962).

SIMULATION

A model is only of real value once it has been tested and proved to behave in a manner similar to that of the system which it is supposed to represent.   A working model is said to "simulate" the system it represents.   Computers which solve mathematical expressions describing the system are called "simulators". Simulation combines the advantages of both mathematical and physical modelling.

A simulator provides an artificial reality for testing a model (Kac, 1969).   A great many simulated experiments can be performed in a short time.   The response to a change in any parameter or set of control characteristics may be examined.   The model can be subjected to extreme stresses which are not possible under experimental conditions with the biological system.   The results of simulation can indicate what biological experimentation is necessary.

The suggestion that analogue simulation might be applied to human heat flow seems to have originated at the U.S. Naval Air Development Centre in about 1954 (Toll, 1954; Hardy, 1953/4).   The first successful analogue computer simulation of human thermoregulation was reported by Wyndham and Atkins in 1960 (Wyndham, 1960) and was followed immediately by that of Crosbie, Hardy and Fessenden (1961, 1963).

In discussing simulation of the thermoregulatory system it is useful to distinguish between the controlled (or passive) and the controlling (or active) parts of the system (Figure 2).   The controlled system is the body of the animal, which may suffer a thermal disturbance.   Models of the controlled system are concerned with the heat flow and temperature distribution within the body.   The controlling system consists of those parts of the whole system which begin by sensing the controlled variable and eventually provide the necessary control action.

Simulation in thermoregulation consists firstly of reducing the biological controlled system to a simple model.   The differential equations governing heat flow in the model are derived.   A mechanism for the controlling system is postulated and described in mathematical form.   The complete set of equations is then solved on a computer.   The computer may be an electrical analogue (or differential analyzer) or a digital computer.   The point must be made that solving equations describing a thermal model on an analogue computer is not the same process as making an electrical model.

The controlled system

The heat flow in a volume element of the body can be described

quantitatively by well-established physical laws in the form of the
diffusion equation. Theoretically, a perfectly general set of
equations of heat flow in the body can be built by using infinitesmal
volume elements to each of which the diffusion equation is applied.
Brown (1963, 1965) has demonstrated how this procedure can be
executed using vector algebra.   In practice, however, "lumped
parameter" models are usually employed, in which the body is divided
into a number of finite layers.   Each layer is assumed to have a
particular structure and set of thermal properties.

     Although some work was done on simple thermal models of the
controlled system during the 1940's and 50's (Pennes, 1948; Taylor,
1956; Friedmann, 1953), it was only 12 years ago that satisfactory
models of the controlled system began to appear.   Wissler (1959,
1961, 1963) reported the solution on a digital computer of
mathematical equations describing the steady state, and later the
transient state, temperature distribution in the nude human body.

     Wissler originally divided his man model into six cylinders:
head, trunk, two arms and two legs.   Each cylinder consisted of one
uniform layer.   Atkins (Atkins, 1962; Atkins, 1963; Wyndham, 1960)
constructed a model in which the body was represented by a single
cylinder containing four concentric layers.   His model is shown in
Figure 4.   The four concentric layers correspond to four different
types of tissue: core, including skeleton, viscera etc.; muscle;
deep skin and superficial fascia; and the bloodless layer of the
skin.   He also included an independent central blood pool which

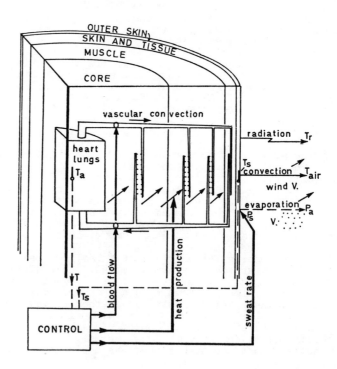

Fig. 4.   Thermal model of human body. (From A.R. Atkins
          and C.H. Wyndham, (1969), Arch. Ges. Physiol.
          307, 104).

collected blood returning from different areas, mixed it in the
heart and lungs, and redistributed it.

Atkins assumed there to be no axial heat flow in his cylinders so
that his model was essentially one-dimensional. It could equally
well have been represented by a semi-infinite slab. The thermal
model employed by Crosbie, Hardy and Fessenden (1961, 1963) was
indeed a multi-layered semi-infinite slab. The concept of a one-
dimensional layered model, either a cylinder or a slab, has subsequently been in general use (Stolwijk, 1968a; Atkins, 1969;
Stolwijk, 1966; Smith, 1964).

The mathematical solutions of the diffusion equation for a
number of different thermal models have been gathered together
recently in a detailed review by Hsu in his doctoral dissertation
(Hsu, 1971) and will not be elaborated here. However we should like
to contribute the following remarks.

Firstly, the equation governing temperature distribution within
the body is a second order differential equation in two variables:
time and a spatial coordinate. It cannot be solved analytically
but only by finite difference techniques (Wissler, 1970). The more
steps included in the solution, the more accurate is the final
temperature distribution. In a layered model, each layer can be
subdivided. The solution may be performed on either an analogue or
a digital computer. There are still some advantages in using
analogue computers: they are generally faster than digital computers
and their operating costs are lower. Digital computers have the
great advantage of versatility and flexibility, and with the advent
of large fast time-sharing machines and the availability of suitable
software such as the IBM Continuous System Modelling Program (CSMP).
they will be used more and more. The useful ness of digital
computers has been greatly enhanced by the discovery in the West of
the "alternating direction" finite difference technique of Saul'ev
(Saul'ev, 1964), which was published in the USSR in 1957. The
Saul'ev technique is fast, accurate and unconditionally stable.

Secondly the equations of temperature distribution are solved
subject to two boundary conditions, namely that the temperature
gradient at centres of symmetry should vanish and that the heat flow
at the surface should equal the heat transfer to the environment.
Since the equations of heat transfer to the environment are as yet
known only for the body as a whole, there is little point in
dividing the body into various segments rather than considering it
as a single cylinder or slab.

Thirdly, thermal properties and local rates of heat generation
have to be assigned to each tissue layer in the model. Stolwijk
and Hardy (1966) have put a great deal of work into collecting what
data are available in the literature but the available data is
insufficient.

Lastly, current methods of modelling heat transfer between the
bloodstream and the surrounding tissue are not satisfactory. It
is usual practice to calculate the heat transfer between blood and
tissue by assuming that the blood enters a tissue element at
arterial blood temperature and leaves at local tissue temperature.
The heat flow so calculated is multiplied by a factor between 0 and
1 to account for incomplete heat transfer and counter-current heat
exchange between large arteries and veins. The value of this
factor is obtained by trial-and-error: values are substituted
until the resulting temperature distribution corresponds with
experimental observations (Stolwijk, 1966). This empirical approach
is hard to justify. Latest evidence shows that in fact there is
rarely counter-current heat exchange between arteries and veins
(J.W. Mitchell, 1968; J.W. Mitchell, 1970). Many models assume
that blood flowing to a particular tissue layer, say the muscles,

all returns to the heart before reaching another layer, say the skin.
In exercising men there appears to be some direct flow from muscle to
overlying skin (Cooper, 1959).   Some models have included this so-
called "squeeze flow" (Smith, 1964) but its quantitative contribution
to total heat transfer is not known.   Finally with regard to
modelling the role of the circulation in heat transfer, the blood
flow rate through a particular tissue layer is generally regarded as
the most important factor, governing the heat transfer rate. Recent
work on the simulation of the responses of exercising men (Atkins,
1971) suggests that in the skin the blood flow distribution within a
layer is equally important.

The controlling system

In spite of the reservations expressed above, models of the
controlled system have advanced much further than those of the
controlling system.   This does not mean that there are no operating
models of the controlling system.   There are several, some of which
simulate actual experimental observations well.   But to quote the
editor of Nature (Editor, Nature, 1971) "it is not difficult to make
models; the difficulty lies in finding the right one".

Let us return to the simple block diagram of a negative feedback
control system (Figure 2).   A model of the controlling system should
ultimately identify the controlled variable, model the behaviour of
the feedback elements, simulate the action of the control elements
and provide a suitable effect on the controlled system.   In other
words, the correct transfer function must be found for each of the
blocks in the block diagram, and for any additional blocks which
need to be introduced.   We are nowhere near able to do this yet.

The fact that operating models of the controlling system do exist
results from liberal use of the "black box" concept (Hardy, 1965).
By experimentation it is possible to derive functional relationships
between control actions and particular body temperatures, without
ever knowing the relationship between those temperatures and the
controlled variables.   One reduces the block diagram to the form
shown in Figure 5, with the black box hiding a host of unknowns.

We have just such a simulator operating very well in our
laboratory.   The simulator, was designed primarily to predict the
reactions of men exposed to combinations of heat and work stress.

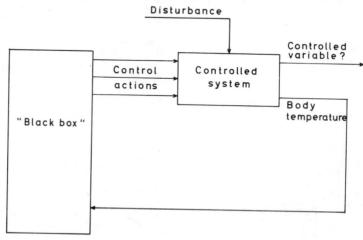

Fig. 5.   "Black box" controlling system.

(Atkins, 1962).   It was not necessary to model the mechanisms of the
controlling system as long as the relationship between a disturbance
and particular physiological strain parameters was correctly predicted.   Into the simulator are built function generators which mimic
the relationship between deep body temperature and skin temperature
on the one hand and sweating and whole body conductance on the other.
Figure 6 shows some of the results generated by the simulator.
The bottom half of the figure shows, as a function of time, the
simulated responses of core temperature and mean skin temperature of
nude resting men after moving from a neutral environment to dry, low
wind-speed environments with various temperatures.   The top half
of the figure shows the rectal temperature and skin temperature of
two men subjected to just those experimental conditions in the wind
tunnel at the Human Sciences Laboratory.   The agreement is good.
Notice for instance the initial overshoot of skin temperature in the
heat, the absence of steady state after two hours in environments of
13°C and 24°C, the virtual independence of skin temperature of air
temperature in hot environments and the quantitative agreement in

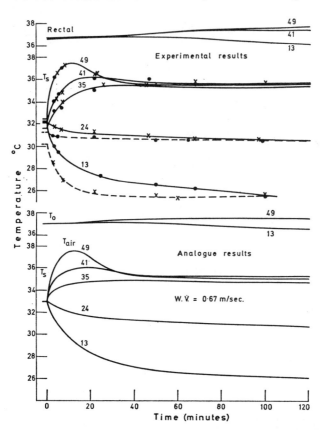

Fig. 6.   Simulation of responses of nude resting men to
           various environments.   (From A.R. Atkins and
           C.H. Wyndham, (1969), Arch. Ges. Physiol. 307, 104).

both transient and steady state responses.

Models which produce the correct results without necessarily simulating the biological mechanisms might be called synthetic. Analytical models, those which do simulate the biological mechanisms, are of much greater fundamental value and are, of course, much more difficult to construct. Even models which are primarily synthetic can serve analytic functions. For example, when we tried to simulate the reactions of a working man using the same black box we used for a resting man, the predicted body temperature and time constants differed considerably from those actually observed. To reproduce the experimental results we needed to increase the sensitivity of the sweating and circulatory systems to changes in body temperature. The suggestion arises that the man might do the same.

A recent model from the Pierce Laboratory (Stolwijk, 1968a; Stolwijk, 1970) has laid the foundation of what might lead to a successful analytic model. Their model of the controlling system, programmed in Fortran, has a degree of flexibility and versatility not previously found. Temperature and its rate of change can be measured in any layer of any segment of the model of the controlled system. The resulting information can be integrated in any desired way. A command signal can then be sent again to any layer of any segment where its effect can be modulated by the local conditions. The advantage of a flexible model is that new experimental results and new hypotheses may be incorporated with very little trouble.

The current approach to modelling the controlling system consists of attempting, on the basis of rather incomplete experimental results, to write equations expressing control actions in terms of combination of body temperatures and their derivatives. The result has been a plethora of empirical equations in which various body temperatures add and multiply in different ways. We agree with Brown and Brengelmann (1970) that the time is not yet ripe to write this kind of equation. None of the attempts has been particularly successful even in modelling the responses of resting man. Wyndham and Atkins (1968) have demonstrated the hazards of trying to write a mathematical equation for a control action as a function of body temperatures.

As Riggs (1963) has pointed out, the fault lies not in the mathematics but in the model to which the mathematics is applied. We believe that the failure of models of the controlling system has three origins. These are firstly a tendency to build models according to theories of thermoregulation which are not necessarily correct and are certainly not the only possibility; secondly, inadequacy of physiological information; thirdly, failure to recognise the importance of available physiological information. The second of these needs no elaboration. Let us discuss the other two.

Constraining models to fit theories

A number of specific examples may be presented. Firstly, almost without exception models of thermoregulation try to explain the interaction of afferent signals from two populations of temperature sensors: one in the core of the body and another in the periphery. Although there are undoubtedly temperature sensors in the core and periphery, and they do interact, there is growing evidence that temperature changes in most, if not all, parts of the body will excite thermoregulatory responses. There is little reason to believe that the thermoregulatory integrating centres treat a thermal afferent signal in a particular way because it arises from a particular region. Except for the integrative parts of the central nervous system, and perhaps the scrotum of the ram, the thermoregulatory response to a thermal disturbance is independent of where in the body the disturbance occurs (J.W. Snellen and D. Mitchell, unpublished results).

Snellen (1966) found a high correlation in men between change in

average body temperature calculated calorimetrically and sweat rate.
We have found (Mitchell, 1971) that in man average body temperature
correlates well not only with sweat rate but also with conductance,
and metabolic rate during shivering. It may well be that the controlled variable is simply average body temperature, an idea supported
recently by Colin and Houdas (1966, 1968), Nielsen (1969) and Brück
Wünnenburg (1967). Neural information relaying average body temperature could be provided by simply adding the information from sensors
uniformly distributed throughout the body. The early analogue simulator of Crosbie, Hardy and Fessenden (1961) used average body temperature as the controlled variable and operated quite successfully.
     We are not yet in a position to say with certainty that average
body temperature is the controlled variable. Attempts to identify
the controlled variable must continue. The point needs to be made
that the only way of identifying the controlled variable in the
thermoregulatory system is to identify the source of feedback. The
alternative, identifying the variable in which variations appear to
be controlled, is not conclusive. If average body temperature is
indeed the controlled variable, many specific body temperatures will
be constant in the steady state. One often encounters the argument
that body core temperature is the most stable of the body temperatures
and is therefore the controlled variable. The argument is illogical.
Everyone is familiar with the technique of producing a very stable
temperature by placing a vaccuum flask in a temperature-controlled
water bath (Figure 7). The temperature inside the flask is the most
stable but the water temperature is controlled. In general when a
heat source in a variable environment is surrounded by thermal resistance, the core temperature will be the most stable whatever temperature in controlled.
     The second example of an perhaps unjustified theoretical constr-
aint on many models concerns the role of the brain and spinal cord.
There is a great deal of evidence for high thermo-sensitivity in the
hypothalamus and to a lesser extent in other parts of the brain and
in the spinal cord. Temperature sensitive neurones have been found
in these regions. As a result brain and spinal cord temperature are
commonly considered to be important afferent signals to the thermoregulatory controller. There is no evidence, however, that there are
a high number of primary temperature receptors in the brain and spinal
cord. That the neurone is highly temperature sensitive does not imply
that it is a thermoreceptor: some neurones with Q_{10} as high a 8
appear to have nothing to do with thermoregulation (Barker, 1970).
Most of the neurones may be interneurones and the role of neural
structures in the brain and spinal cord may be that of amplification,
the gain of which depends on the local temperature. It seems as
though thermoregulatory responses can be elicited by heating and
cooling any part of the central nervous system containing synapses
in thermoregulatory pathways (Cabanac, 1969; Jessen, 1971).

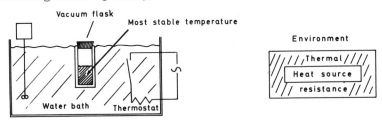

Fig. 7. Demonstration that the most stable temperature is not
         necessarily the controlled temperature.

Differences in sensitivity may be the result of differences in synapse
density.

The third theory entrenched in many models of thermoregulation is
that of a set-point provided by a reference signal. This stems from
a tendency to think of biological control systems in terms of engineering components (D. Mitchell, 1970). There is little reason to
prefer a reference signal theory to the type of dynamic balance control originally proposed by Bazett (1927) and demonstrated in Figure
8. It is quite possible to have an apparent set-point without there
being any reference signal (D. Mitchell, 1970). Models of thermoregulation have been shown to operate without reference signals (Cornew,
1967; Smith, 1964). Our analogue simulator incorporates no reference
signal (Atkins, 1969).

Failure to recognise important information

Again, we shall discuss specific examples of situations where
physiological information is available but it has not been recognised
as important, has been misinterpreted or has simply not been included
in models. The first and most obvious example concerns the direct
influence of local temperature on thermoregulatory control actions.

The effect of local temperature on local sweat production is well
documented. It is generally accepted that the mechanism is a modulation at the neuroglandular junction of the effect that the neural
stimulus has on the sweat gland (MacIntyre, 1968; Ogawa, 1970). The
magnitude of the effect has not been generally realised. Bullard
(1970a) has reported that local heating enhances the response of the
sweat gland to stimulation with a maximum Q_{10} of 5 - 6. Local
skin temperature is at least as important as any other body temperature in the control of sweating.

Only recently (Nadel, 1971; Bullard, 1970b) has the role of the
local modulation of sweating by local temperature been recognised and
built into a model. Local skin temperature acts as a multiplier
operating on an efferent signal from a central controller. Skin
temperature must necessarily enter into an equation describing the
control of sweating as a multiplicative term whether or not neural
sensors of skin temperature are involved in the control of sweating.

In an analogous way local temperature seems to modulate the
effect of efferent neural signals from a central controller on the

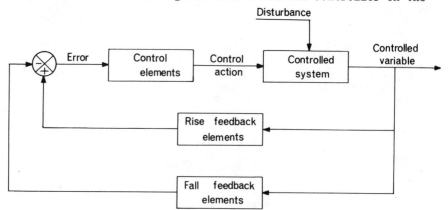

Fig. 8.   Dynamic balance negative feedback control. (From
          D. Mitchell, J.W. Snellen and A.R. Atkins, (1970),
          Arch. Ges. Physiol. <u>321</u>, 323).

smooth muscle of blood vessels (Hertzman, 1963; Webb-Peploe, 1968; Rowell, 1971; Davies, 1971). The magnitude of the effect has not been measured and no model includes it. In addition, local temperature also seems to influence the contraction of the smooth muscle of blood vessels in a way which does not depend on arriving neural signals at all (Hertzman, 1963; Walther, 1971; Bligh, 1971).

The second example of available physiological information not exploited in models concerns the firing characteristics of temperature-sensitive neurones. Much detailed information has been gathered in the last decade on the firing characteristics of peripheral temperature-sensitive neurones over a wide temperature range, particularly in the laboratories of Drs. Hensel, Iggo and Kenshalo. These firing patterns are important to modellers because they determine the nature of the feedback. It must be more than coincidence that temperature-sensitive neurones show strong dynamic responses (Hensel, 1960) to rates of change of their temperature, and that certain control actions are generated, at least in part, by rates of change of temperature (Hardy, 1961; Banerjee, 1969; Wurster, 1969; Nadel, 1971).

Last year in Lyon, Drs. Guieu and Hardy (Hardy, 1970) reported their attempts to build neurone firing characteristics into models. One hopes the work will continue and expand.

The third example of the experimentalists outpacing the modellers lies in the field of exercise in man. Much research has been performed in recent years on thermoregulation during exercise, but successful models of thermoregulation during exercise have been noticeably absent. One of the most important experimental results has been the identification, largely through work in Dr. Robinson's laboratory (Robinson, 1965; Gisolfi, 1971) of a non-thermal factor involved in the control of sweating during exercise. We mentioned earlier that computer simulation suggested that the sensitivity of both the sudomotor and vasomotor control systems increased in exercise. We now have direct experimental evidence for an increase in gain during exercise, probably humorally mediated. The evidence arises from a series of experiments in the wind tunnel at the Human Sciences Laboratory, reported in detail elsewhere (Mitchell, 1971). In a number of different environments two heat-acclimatized nude men sat on the bicycle ergometer for 90 minutes, then worked continuously at one of two levels for 90 minutes and finally recovered at rest for 90 minutes. The men were in thermal equilibrium, as measured by direct calorimetry, at the end of the rest, exercise and recovery periods.

Two significant results emerged. Firstly, during the exercise period, the increase in body temperature (mean body, oesophageal, and rectal) for a given increase in sweat rate depended on the environmental temperature. The cooler the environment, the greater the increase. In the cooler environments the increase tended to be more than that observed in resting men for the same increase in sweat rate, in hotter environments less. In the hottest environments body temperature was sometimes even lower in exercise than in rest. Secondly, body temperatures remained elevated after exercise, relative to pre-exercise levels, even though the men were in thermal equilibrium and sweat rates and conductances returned to pre-exercise levels. The post-exercise elevation was linearly correlated with the elevation during exercise, and was not a result of dehydration. The second significant result, therefore, put simply, is that it requires a much smaller change in temperature to decrease sweating and conductance a given amount after exercise than to increase them by that amount during exercise. As far as we are aware, these observations have only one explanation: the gain of the thermoregulatory system increases during exercise, and the high gain persists for some time after the exercise.

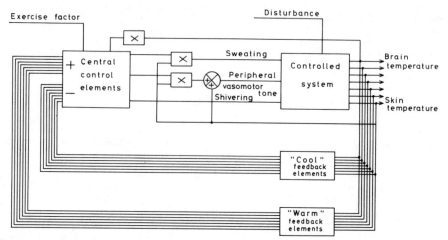

Fig. 9. Tentative block diagram for short-term thermoregulation
        in man.

Haight and Keatinge (1969) have noticed the elevation of body temperature after exercise and report the elevation to persist for at
least 5 hours, suggesting a humoral rather than a neural factor.
Bodil Nielsen (1968) has suggested that body temperature levels in
exercise are influenced by chemical factors released during the
exercise. The factor has not yet been identified. It is something
that appears quickly and disappears slowly. Because it is a humoral
factor it is unlikely to cause the extremely rapid increase in sweating at the onset of work noticed by van Beaumont and Bullard (1966).
    To conclude the section on the controlling system, we should like
to present a tentative block diagram for short-term thermoregulation
in man incorporating some of the concepts just discussed (Figure 9).
Its important features are the absence of a reference signal, mean
body temperature as the controlled variable, control of the gain of
the integrating centre by both brain temperature and exercise, local
modulation of sweating and peripheral vasomotor tone by local temperature and a direct influence of local temperature on peripheral vasomotor tone. The model also displays some characteristics of parallel
path feedback control, which Brown and Brengelmann (1970) have
observed to occur in almost all biological control systems, adding to
reliability and stability.

CONCLUSION
    As Hardy and Stolwijk (1968) have pointed out, models are never
correct. The model we have just presented is not correct - it is
something to work on. If it stimulates research to prove it wrong,
it has achieved its purpose.
    Modelling physiological phenomena without a simultaneous experi-
mental programme is extremely hazardous and all too common. The
editor of Bio-Medical Engineering (1971) puts it well: "reducing a
biological system to a physical system and then analysing it squeezes
out all the biology that exists". Models need to be challenged.
Probably the best way of challenging a model of a control system is to
compare its predictions with actual observations during cyclical
application of a disturbance (Kastella and Brown, 1970; Piironen,
1970).
    Finally, a quotation from Kac (1969): "models are, for the most
part, caricatures of reality, but if they are good, then, like good
caricatures, they portray, though perhaps in a distorted manner, some

of the features of the real world". Models can become "reality" –
atoms and genes were originally only components of models.

ACKNOWLEDGEMENTS

This paper constitutes part of the research effort of the Chamber
of Mines of South Africa.
We are grateful to Mr. D. Rabe for preparation of the figures, to
Dr. E.R. Nadel for a preprint of unpublished work, and to Dr. J.W.
Snellen for many useful discussions about thermoregulation.

REFERENCES

Atkins, A.R., (1962), A method of simulating heat flow and control in
a nude man with an analogue computer, M. Sc. (Eng.) dissertation,
University of the Witwatersrand, Johannesburg, South Africa.
Atkins, A.R., (1963), An electrical analogue of heat flow in the
human body, South African Mech. Engr. $\underline{13}$, 40.
Atkins, A.R., and C.H. Wyndham, (1969), A study of temperature
regulation in the human body with the aid of an analogue computer,
Arch. Ges. Physiol. $\underline{307}$, 104.
Atkins, A.R., and D. Mitchell, (1971), Analogue computer simulation
of the thermal response of a working man, International symposium
on Environmental Physiology : Bioenergetics and Temperature
Regulation, Dublin.
Banerjee, M.R., R. Elizondo, and R.W. Bullard, (1969), Reflex
responses of human sweat glands to different rates of skin cooling,
J. Appl. Physiol. $\underline{26}$, 787.
Barker, J.L., and D.O. Carpenter, (1970), Thermosensitivity of
neurones in the sensorimotor cortex of the cat, Science $\underline{196}$, 597.
Bazett, H.C., (1927), Physiological responses to heat, Physiol. Rev.
$\underline{7}$, 583.
Bligh, J., W.H. Cottle, and M. Maskrey, (1971), Influence of ambient
temperature on the thermoregulatory responses to 5-hydroxytrypta-
mine noradrenaline and actycholine injected into the lateral
cerebral ventricles of sheep, goats and rabbits, J. Physiol.
(London) $\underline{212}$, 377.
Brown, A.C., (1963), Analog computer simulation of temperature
regulation in man, Report AMRL-TDR-63-116, Biomedical Laboratory,
6570th Aerospace Medical Research Laboratories, Wright-Patterson
Air Force Base, Ohio.
Brown, A.C., (1965), Equations of heat distribution within the body,
Bull. Math. Biophys. $\underline{27}$, 67.
Brown, A.C., and G.L. Brengelman, (1970), The interaction of
peripheral and central inputs in the temperature regulation system,
In : Physiological and Behavioural Temperature Regulation, Editors:
J.D. Hardy, A.P. Gagge, and J.A.J. Stolwijk, (C.C. Thomas).
Brück, K., and W. Wünnenberg, (1967), Die Steuerung des Kältezitterns
beim Meerschweinchen, Arch. Ges. Physiol. $\underline{293}$, 215.
Bullard, R.W., (1970a), Electrical stimulation of human sweat glands
and local heating enhancement, Physiologist 13, 158.
Bullard, R.W., M.R. Banerjee, F. Cohen, R. Elizondo, and B.A.
MacIntyre, (1970b), Skin temperature and thermoregulatory sweating:
a control systems approach, In : Physiological and Behavioural
Temperature Regulation, Editors : J.D. Hardy, A.P. Gagge, and
J.A.J. Stolwijk, (C.C. Thomas).
Cabanac, M., and J.D. Hardy, (1969), Responses unitaires et
thermoregulatrices lors de rechauffements et refroidissements
localises de la region preoptique et du mesenchephale chez le
lapin, J. Physiol. (Paris) $\underline{61}$, 331.
Colin, J., and Y. Houdas, (1966), La notion de temperature moyenne
du corps dans l'etude du declenchement de la sudation thermique,
Compt. Rend. Soc. Biol. $\underline{160}$, 2076.

Colin, J., and Y. Houdas, (1968), Determinisme du declenchement de la
    sudation thermique chez l'homme, J. Physiol. (Paris) 60, 5.
Cooper, T., W.C. Randall, and A.B. Hertzman, (1959), Vascular
    convection of heat from active muscle to overlaying skin, J. Appl.
    Physiol. 14, 207.
Cornew, R.W., J.C. Houk, and L. Stark, (1967), Fine control in the
    human temperature regulation system, J. Theoret. Biol. 16, 406.
Crosbie, R.J., J.D. Hardy, and E. Fessenden, (1961), Electrical
    analogue simulation of temperature regulation in man, Report
    NADC-MA-6130, U.S. Naval Air Development Center, Johnsville.
Crosbie, R.J., J.D. Hardy, and E. Fessenden, (1963), Electrical
    analog simulation of temperature regulation in man, In : Temper-
    ature : Its Measurement and Control in Science and Industry, Vol. 3,
    Editor-in-Chief : C.M. Herzfeld, Part 3, Biology and Medicine,
    Editor : J.D. Hardy, (Reinhold).
Dainty, J., (1960), Electrical analogues in biology, In : Symposia of
    the Society for Experimental Biology No. 24, Models and Analogues
    in Biology.
Davies, B.N., D.A. Powis, and P.G. Withrington, (1971), Effects of
    lowered temperature on the responses of the smooth muscle of the
    isolated, blood-perfused dog's spleen to sympathetic nerve stimu-
    lation, J. Physiol. (London) 212, 18P.
Editor, Bio-Medical Engineering, (1971), Engineers in health care,
    Bio-Med. Eng. 6, 105.
Editor, Nature, (1971) Model biology, Nature 229, 87.
Friedman, N.E., and K. Buettner, (1953), Report on biothermal
    analyzer, Memorandum Report No. 2, Thermal Biotechnology Project,
    Dept. of Engineering, U.C.L.A.
Gisolfi, C., and S. Robinson, (1970), Central and peripheral stimuli
    regulating sweating during intermittent work in men, J. Appl.
    Physiol. 29, 761.
Haight, J.S.J., and W.R. Keatinge, (1969) Human temperature
    regulation afterprolonged intermittent exercise, J. Physiol
    (London), 206, 20P.
Hammel, H.T., (1968), Regulation of internal body temperature, Ann.
    Rev. Physiol. 30, 641.
Hardy, J.D., (1953/4), Control of heat loss and heat production in
    physiologic temperature regulation, Harvey Lectures, Series 49,249.
Hardy, J.D., (1961), Physiology of temperature regulation, Physiol.
    Rev. 41, 521.
Hardy, J.D., (1965), The 'set-point' concept in physiological temp-
    erature regulation, In : Physiological Controls and Regulations,
    Editors : W.S. Yamamoto and J.R. Brobeck, (W.B. Saunders).
Hardy, J.D., and J.A.J. Stolwijk, (1968), Regulation and control in
    physiology, In : Medical Physiology, Editor : V.B. Mountcastle,
    Vol. i, 12th edition, (C.V. Mosby).
Hardy, J.D., and J.D. Guieu, (1970), Integrative activity of preoptic
    units II : An hyperthetical interneuronal network for temperature,
    Sumposium Internationale de Thermoregulation Comportementale,Lyon.
Hensel, H., A. Iggo, and I. Witt, (1960), A quantitative study of
    sensitive cutaneous thermoreceptors with C afferent fibres, J.
    Physiol. (London) 153, 113.
Hertzman, A.B., (1963), Regulation of cutaneous circulation during
    body heating, In : Temperature : Its Measurement and Control in
    Science and Industry, Vol. 3, Editor-in-Chief : C.M. Herzfeld,
    Part 3, Biology and Medicine, Editor : J.D. Hardy, (Reinhold).
Hsu, F., (1971), Modelling, simulation, and optimal control of the
    human thermal system, Ph.D. dissertation, Kansas State University,
    Manhattan.
Jessen, C., and E.T. Mayer, (1971), Spinal cord and hypothalamus as
    core sensors of temperature in the conscious dog I Equivalence of

responses, Arch. Ges. Physiol. 324, 189.

Kac, M., (1969), Some mathematical models in science, Science 166,695.

Kastella, K.G., and A.C. Brown, (1970), Effect of hypothalamic temperature waveforms on peripheral blood flow in the baboon, J. Appl. Physiol. 29, 499.

MacDonald, D.K.C., and C.H. Wyndham, (1950), Heat transfer in man, J. Appl. Physiol. 3, 342.

MacIntyre, B.A., R.W. Bullard, M. Banerjee, and R. Elizondo, (1968), Mechanism of enhancement of eccrine sweating by localized heating, J. Appl. Physiol. 25, 255.

Mitchell, D., J.W. Snellen, and A.R. Atkins, (1970), Thermoregulation during fever : change of set-point or change of gain, Arch. Ges. Physiol. 321, 293.

Mitchell, D., (1971), Human surface temperature : Its measurement and its significance in thermoregulation, Ph.D. thesis, University of the Witwatersrand, Johannesburg.

Mitchell, J.W., and G.E. Myers, (1968), An analytical model of the counter-current heat exchange phenomena, Biophysics J. 8, 897.

Mitchell, J.W., T.L. Galvez, J. Hengle, G.E. Myers, and K.L. Siebecker (1970), Thermal response of human legs during cooling, J. Appl. Physiol. 29, 859.

Nadel, E.R., R.W. Bullard, and J.A.J. Stolwijk, (1971), Importance of skin temperature in the regulation of sweating, J. Appl. Physiol. 31(1).

Nielsen, B., (1968), Thermoregulatory responses to arm work, leg work and intermittent leg work, Acta Physiol. Scand. 72, 25.

Nielsen, B., (1969), Thermoregulation in rest and exercise, Acta Physiol. Scand. 79, suppl. 323.

Ogawa, T., and R.W. Bullard, (1970), Local effect of skin temperature sudorific agents, Physiologist 13, 265.

Pennes, H.H., (1948), Analysis of tissue and arterial blood temperatures in the resting human forearm, J. Appl. Physiol. 1, 93.

Piironen, P.P., (1970), Sinusoidal signals in the analysis of heat transfer in the body, In : Physiological and Behavioural Temperature Regulation, Editors : J.D. Hardy, A.P. Gagge, and J.A.J. Stolwijk, (C.C. Thomas).

Riggs, D.S., (1963), The Mathematical Approach to Physiological Problems, (Williams and Wilkins).

Robinson, S., F.R. Meyer, J.L. Newton, C.H. Ts'ao, and L.O. Holgersen, (1965), Relations between sweating, cutaneous blood flow and body temperature in work, J. Appl. Physiol. 20, 575.

Rowell, L.B., G.L. Brengelmann, J-M, R. Detry, and C. Wyss, (1971), Venomotor responses to local and remote thermal stimuli to skin in exercising man, J. Appl. Physiol. 30, 72.

Saul'ev, V.K., (1964), Integration of Equations of Parabolic Form by the Method of Nets, (Pergamon).

Smith, P.E., and E.W. James, (1964), Human responses to heat stress, Arch. Environ. Health 9, 332.

Snellen, J.W., (1966), Mean body temperature and the control of thermal sweating, Acta. Physiol. Pharmacol. Neerl. 14, 99.

Stolwijk, J.A.J., and J.D. Hardy, (1966), Temperature regulation in man - a theoretical study, Arch. Ges. Physiol. 291, 129.

Stolwijk, J.A.J., and D.J. Cunningham, (1968a), Expansion of a mathematical model of thermoregulation to include high metabolic rates, Final Report - A NAS-9-7140, John B. Pierce Foundation Laboratory, New Haven.

Stolwijk, J.A.J., B. Saltin, and A.P. Gagge, (1968b), Physiological factors associated with sweating during exercise, Aerospace Med. 39, 1101.

Stolwijk, J.A.J., (1970), Mathematical model of thermoregulation, In: Physiological and Behavioural Temperature Regulation, Editors :

J.D. Hardy, A.P. Gagge, and J.A.J. Stolwijk, (C.C. Thomas).
Taylor, C.L., (1956), Descriptions and predictions of human response
    to aircraft thermal environments, Presented at ASME Aviation
    Conference, Cited by : Woodcock, A.H., H.L. Thwaites and J.R.
    Breckenridge, An electrical analogue for studying heat transfer in
    dynamic situations, Technical Report EP - 86, U.S. Headquarters,
    Quartermaster Research and Engineering Command, Natick.
Toll, E. (1954), Temperature response of the human body to thermal
    stimuli- analogue comouter simulation, Report Note 46-54, U.S.
    Naval Air Development Center, Johnsville.
Van Beaumont, W., and R.W. Bullard, (1966), Sweating exercise
    stimulation during circulatory arrest, Science 152, 1521.
Walther, O-E, E. Simon, and C. Jessen (1971), Thermoregulatory adjust-
    ments of skin blood flow in chronically spinalized dogs, Arch. Ges.
    Physiol. 322, 323.
Webb-Peploe, M.M., and J.T. Shepherd, (1968), Peripheral mechanism
    involved in response of dogs' cutaneous veins to local temperature
    change, Circulation Res. 23, 701.
Wissler, E.H., (1959), Steady state temperature distribution in the
    human, Report No. 1, Contract DA-49-007-2005, University of Texas,
    Austin.
Wissler, E.H., (1961), An analysis of factors affecting temperature
    levels in the nude human, Report No. 4, Contract DA-49-007-MD-2005,
    University of Texas, Austin.
Wissler, E.H., (1963), An analysis of factors affecting temperature
    levels in the nude human. In : Temperature : Its Measurement and
    Control in Science in Industry, Vol. 3., Editor-in-Chief: C.M.
    Herzfeld, Part 3, Biology and Medicine, Editor : J.D. Hardy,
    (Reinhold).
Wissler, E.H., (1970), The use of finite difference techniques in
    simulating the human thermal system, In : Physiological Behavioural
    Temperature Regulation, Editors: J.D. Hardy, A.P. Gagge, and J.A.J.
    Stolwijk, (C.C. Thomas).
Wurster, R.D., and R.D. McCook, (1969), Influence of rate of change in
    skin temperature on sweating, J. Appl. Physiol. 27, 237.
Wyndham, C.H., W. v.d.M. Bouwer, M.G. Devine, H.E. Paterson, and
    D.K.C. MacDonald, (1952), Examination of use of heat-exchange
    equations for determining changes in body temperature, J. Appl.
    Physiol. 5, 299.
Wyndham, C.H., and A.R. Atkins, (1960), An approach to the solution
    to the human biothermal problem with the aid of an analogue
    computer, Proc. 3rd International Conference on Medical Electronics
    London.
Wyndham, C.H., and A.R. Atkins, (1968). A physiological scheme and
    mathematical model of temperature regulation in man, Arch. Ges.
    Physiol. 303, 14.

UNIT ACTIVITY STUDIES OF THERMORESPONSIVE NEURONS

Joseph S. Eisenman
Mt. Sinai Medical School
New York, N.Y., U.S.A.

One of the most powerful tools of present-day neurophysiology is the micro-electrode technique for recording the activity of single cells, or units, in the central nervous system. Utilizing this method, great advances have been made in our knowledge of the functional interrelations among neurons in many CNS systems. Inevitably, the technique has been applied to the central neural structures controlling body temperature with hopes of similar, dramatic results. This essay will review some of the work done and indicate possible directions for future study.

Overall, progress in this area has been surprisingly slow and, in one respect, disappointing. One reason for the slow progress is, in fact, inherent in the microelectrode technique. When sampling CNS activity one cell at-a-time, it is necessary to build up an adequate population of cells before attempting to draw general conclusions as to types of responses, distributions of interesting cells, etc. The need for caution is well illustrated by our experiences in the early studies on preoptic area (POA) thermosensitive neurons. In the first experiments (Nakayama, et al., 1961,1963), no cool-sensitive cells were found. That is, cells whose firing rates increased with POA cooling were not seen. Only later, as the population of cells studied expanded, were such neurons found in the dog (Hardy, et al., 1964), in the cat (Eisenman and Jackson,1967) and in the rabbit (Cabanac, et al., 1968).

Another problem inherent in the method is that of determining whether a particular neuron is, in fact, functioning as part of the thermoregulatory system and if it is, what role it is playing. Is it a detector of local temperature, an afferent or an efferent interneuron? Microelectrode studies of spinal reflex organization, or even of somato-sensory and -motor systems are comparatively easy to interpret since inputs and/or outputs can be readily isolated or identified. In addition, the comparatively few synapses in these pathways makes it possible to study them in anesthetized preparations without undue depression of the basic organization. The electrophysiology of the brain core, including the hypothalamus, is far more complex since readily isolated pathways or functional centers do not exist. A hypothalamic neuron may be operating in any of the regulatory or behavioral systems known to be influenced by activity in this area of the brain. Further, the short neuron, multi-synaptic nature of the input and output pathways in the brain core makes them very sensitive to anesthetic depression. The studies on anesthetized animals give us a picture of activity in a severely truncated system and must be viewed accordingly.

One advantage that we have in studying thermoregulation is given us by our ability to activate POA thermodetectors by local heating and cooling. It is generally assumed that changes in neuronal firing induced by thermal stimulation of the POA do represent neuronal correlates of the regulatory responses evoked by this stimulus. When recordings are taken from the area whose temperature is being varied, even this assumption must be applied cautiously, as pointed out by Barker and Carpenter (1970). (But, cf. Eisenman and Edinger (1971) for comment on this report.) When recording activity in other areas influenced by thermal stimulation of POA or other known thermosensitive structures, the presumption that thermoregulatory responses are being studied is, by far, the most reasonable one.

The problem of determining what function a given neuron plays in the regulatory system is a more difficult one and, as will be discussed below, has yet to be solved. Until this can be done, it will be impossible to put all of our information on neuronal activity together into a proper neural control network.

Disappointment in the results obtained to date stems largely from the fact that electrophysiological studies have served mainly to confirm data obtained by other methods, such as thermal stimulation and ablations. Thus, we have confirmed the presence of highly thermosensitive cells (detectors?) in the preoptic area. Inputs

from spinal cord to hypothalamus and convergence of activity in the posterior hypothalamus have likewise been confirmed. The method has, however, provided specific
information on the levels at which interactions or integrations occur. It has also
given at least one new, unexpected result, as will be discussed later.

The earliest application of the single-unit recording technique to the study of
central neural control of body temperature was in the work of Birzis and Hemingway
(1957), who correlated the activity of neurons in the midbrain and posterior hypothalamus with the onset and cessation of shivering in lightly anesthetized cats.
The firing rates of these cells, which were shown to be efferent neurons, decreased
when shivering was inhibited by warming the skin of the animal, even in the absence
of any rise in central (rectal) temperature. In addition to demonstrating the
feasibility of such studies, Birzis and Hemingway made two important observations
that are still being amplified today: firstly, at the diencephalic and midbrain
levels shivering intensity is signaled by changes in the tonic firing rate of the
effector neurons; that is, the phasic or bursty character of shivering appears to
be organized at some, presumably lower, CNS level. Secondly, inputs from cutaneous
thermoreceptors project at least as high as the posterior hypothalamus.

The period of intensive study of the neuronal activity associated with thermoregulation and particularly of preoptic thermosensitivity began with Nakayama's and
my studies in Hardy's laboratory in 1960 (Nakayama, et al., 1961,1963). The changes
in unit firing rates of septal and preoptic neurons induced by heating and cooling
the block of tissue containing those neurons were determined. As shown in fig. 1,
a variety of thermal response curves was found although all had a positive slope;
i.e., all were warm-sensitive cells. In a continuation of these studies, Jackson
and I (Eisenman and Jackson,1967) proposed a scheme to classify the recorded
responses as being derived from either primary thermodetector neurons or from cells
being driven by the detectors and/or by other inputs to the hypothalamus; i.e.,
interneurons in the afferent or efferent thermoregulatory pathways. It was
postulated that the output from a detector cell should be some continuous function

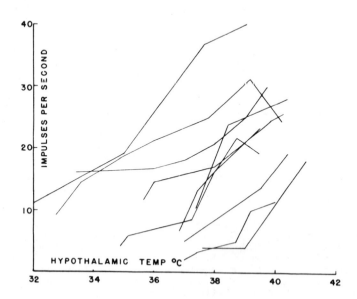

Fig. 1. Thermal response curves from POA neurons, in cats. (From
         Nakayama, et al.,1963)

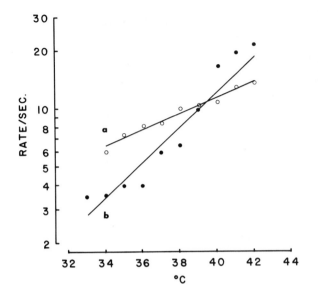

Fig.2.   Thermodetector-
type ($Q_{10} > 2$) response
curves from POA neurons.
(From Eisenman and
Jackson, 1967)

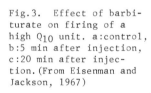

Fig.3.   Effect of barbiturate on firing of a
high Q_{10} unit. a:control,
b:5 min after injection,
c:20 min after injec-
tion.(From Eisenman and
Jackson, 1967)

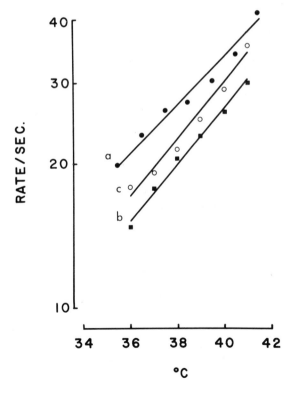

of the input stimulus. That is, firing rate of a thermodetector should change continuously with change in local temperature. In contrast, interneurons might be
expected to show some threshold effects; temperatures above or below which activity
either disappeared or did not change, in keeping with the thresholds for activation
of effector mechanisms. Thermal response curves for interneurons should show dis-
continuities of slope. A further postulate was that a true, primary detector cell
should be spontaneously active due to some inherent membrane mechanism, rather
than being activated by tonic synaptic inputs. This type of mechanism is generally
assumed for peripheral receptors showing tonic or resting activity. Such a mechanism should be relatively insensitive to depression by anesthetic, compared to the
multi-synaptic activation of interneurons. Finally, the work of Magoun, et al.,
(1938) had demonstrated that highest thermosensitivity was localized to the preoptic area at the level of, and between, the anterior commissure and the optic
chiasm.
    We therefore examined the correlation between these three properties of the
recorded units: 1) continuous function thermal response curves, 2) relative insen-
sitivity to anesthetic depression, and 3) histological localization to the preoptic
area, and found that these characteristics did, in fact correlate very well. Those
units with continuous thermal response curves (fig.2) were only minimally depressed
by large, intravenous doses of a short-acting barbiturate, Methohexital Na (fig.3),
and were found primarily in the thermosensitive preoptic area. On the other hand,
units with discontinuous thermal response curves (fig.4) and thermally insensitive
units were markedly depressed by barbiturate and were widely distributed in both
the preoptic area and the septum, as would be expected from interneurons.
    All of the cool-sensitive responses recorded in this study were of the discon-
tinuous type and so were classified as coming from interneurons. Some continuous
function or linear cool-sensitive responses have been reported by Hardy, et al.
(1964) in dogs and by Cabanac, et al. (1968) in rabbits. In a later study on
rabbits (Guieu and Hardy, 1970) no continuous or linear function cool-sensitive

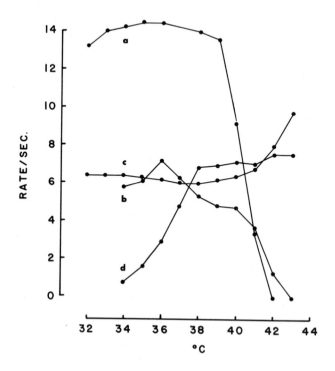

Fig.4. Discontinuous
thermal response
curves: interneuron
types. (From Eisenman
and Jackson, 1967)

POA units were found. If distinct POA cold-detector cells exist, they would appear to be quite rare, or much more difficult to isolate by this technique.

Since the most highly sensitive units with continuous response curves had exponentially increasing firing rates with warming, we adopted a "Q_{10}" terminology for describing these neurons. A plot of log (firing rate) vs. POA temperature could be fitted with a straight line (cf. fig. 2) whose slope was proportional to thermosensitivity. This slope could be expressed as a Q_{10} value; i.e., the ratio of firing rates over an extrapolated 10° interval. The "Thermodetector-type" cells had Q_{10}s greater than 2 ($Q_{10}>2$). Slightly sensitive cells ($Q_{10}2$ or less) and insensitive cells ($Q_{10}1$) were not localized primarily to the preoptic area. In addition, the $Q_{10}1$ units were quite sensitive to barbiturate.

As pointed out by Guieu and Hardy (1970), a Q_{10} designation tends to favor slowly firing units. For example, a cell whose firing rate changes from 2/sec to 8/sec over a 10° range will have a Q_{10} of 4. But one whose rate goes from 20/sec to 80/sec will also be $Q_{10}4$. In terms of firing rate change in spikes/sec·$^{\circ}$C, the latter cell is much more sensitive. The primary reason for choosing Q_{10} to characterize thermosensitivity is that the more sensitive response curves are best fitted by semi-log plots rather than linear ones. In addition, if one can extrapolate from psychophysical measurements to POA detectors, most receptors appear to code intensity as some non-linear, power function. It may well be that ratios of firing rates are more important than absolute firing level.

The bias introduced by using a ratio scale becomes significant only when the firing rates of the neurons being compared differ by orders of magnitude. In these early studies, rates fell for the most part in the range of 2-20/sec, and so were comparable. However, as more variables are studied and firing levels changed, for example, by application of drugs, the problem of how best to characterize a neuron's thermosensitivity may require re-evaluation.

Guieu and Hardy (1970) have classified POA units in rabbits as linear (i.e., having continuous response curves) and non-linear (i.e., having discontinuous response curves). As pointed out earlier, all of the linear responses were of the warm-sensitive type. They also observed that inputs to POA from spinal cord thermosensors ended only on the non-linear type neurons, suggesting an interneuronal role for these cells. Thus, high Q_{10}, $Q_{10}>2$, continuous and linear may all be synonyms for POA thermodetector neurons, while discontinuous or non-linear may be synonyms for interneurons.

Further support for the hypothesis that $Q_{10}>2$ or linear cells are detectors is given by the results obtained by Edinger in my laboratory (Edinger and Eisenman, 1970) on the distributions of thermosensitive cell types in the posterior hypothalamus(table 1). The most striking difference between POA and posterior hypothalamus is the relative scarcity of high Q_{10} or linear cells in the posterior

Table 1. Percentage distribution of units in the preoptic area and posterior hypothalamus.

Neuron Type	Preoptic area	Posterior Hypothalamus
$Q_{10}1$ units	43.2%	47.0%
$Q_{10}2$ units	15.9	17.0
high Q_{10} thermodetectors	21.6	7.0
warm-interneurons	7.9	11.0
cool-interneurons	11.4	16.0
warm-cool units	0.0	2.0

area, as would be expected if this type of response is obtained from detector type neurons.

As indicated in the table, two units were found which fired only at the extreme ends of the thermal range (warm-cool cells), increasing their firing rates with both heating and cooling. It was suggested that these could be a part of a generalized alerting or arousal system responding to potentially harmful levels of central temperature rather than being neurons in a specific thermoregulatory effector pathway. This observation illustrates the need for thinking in terms of the many different central neural functions that may be influenced by changes in central temperature. The possible correlation between thermosensitive neuronal activity and, for example, levels of sleep and wakefulness is just beginning to be investigated (Parmeggiani, et al., 1971). Neurons of the warm-cool type have not been reported in the POA.

Wünnenberg and Hardy (1971) have studied responses of posterior hypothalamic neurons in rabbits to local (i.e., posterior), POA and spinal thermal stimulation. They found neurons responding to various combinations of these inputs: some only to local, posterior hypothalamic temperature changes; some to one or the other of these inputs; some to all three. Both warm-sensitive and cool-sensitive, linear and nonlinear cells were found. The high degree and complexity of convergence in the posterior hypothalamus has thus been demonstrated. This study raises a problem with regard to the classification of cells as detectors or interneurons. Units were found which gave linear or continuous (high Q_{10}) response curves with posterior hypothalamic thermal stimulation. Yet, in contrast to the study of Guieu and Hardy (1970) in the POA, most of these cells also responded to inputs from other areas (spinal cord and/or POA). Do these observations mean that detector cells in the posterior hypothalamus also receive synaptic inputs from other neurons, rather than being first order neurons in a chain, as suggested for the POA detectors? Or do the observations mean that linearity of response is not a good indicator of detector function, at least in the posterior hypothalamus?

Nutik (1971) has also studied posterior hypothalamic neurons, and noted convergence of POA and skin inputs.

Responses of POA units to skin or ambient temperature changes have also been examined. Wit and Wang (1968) reported finding POA cells in anesthetized cats whose firing rate increased when heat was applied peripherally. Hellon (1970) has been studying POA responses to both local and to ambient temperature changes in unanesthetized, chronically prepared rabbits. He finds that some cells are sensitive to both stimuli. Of these, most respond in the same way to POA and peripheral temperature change. This result in the POA contrasts with that of Wünnenberg and Hardy (1971) who found many posterior hypothalamic cells that responded in an inverse way to different inputs; e.g., decreased firing rate with POA heating and increased firing with spinal cord heating. A possible implication is that the posterior hypothalamus is a more complex integrative area than the POA.

The role of the midbrain as a thermosensitive area and as an integrative center for regulatory activity has been investigated by Nakayama and Hardy (1969) and by Nakayama and Hori (1971). Cells in POA with linear response curves to local temperature change were not affected by midbrain thermal stimulation. On the other hand, some POA cells with non-linear (discontinuous) responses were also responsive to midbrain heating or cooling. No midbrain neurons were found to respond to POA thermal stimuli, but many were cool-sensitive with local, midbrain cooling and also responded to skin cooling or to mechanical stimulation of the skin. These responses to peripheral stimuli were abolished by barbiturate anesthetic, as would be expected of inputs over a multi-synaptic pathway.

Some general conclusions can be reached from these studies. In the preoptic area, neurons with high thermosensitivity ($Q_{10}>2$, continuous or linear response) have properties and distributions suggesting that they are thermodetectors. Such cells are not influenced by stimulation of the midbrain or spinal cord, suggesting a pure detector function. Interneuron-type POA cells (discontinuous or non-linear) do respond to midbrain or spinal cord inputs. In the posterior hypothalamus, convergence from POA and spinal cord and from POA and skin has been demonstrated for both detector and interneuron-type cells. Whether this means that some posterior hypothalamic neurons serve a dual role as detectors and interneurons remains to be determined. There is no theoretical reason for rejecting such a dual function.

Very few linear, cool-sensitive responses have been found in the POA. Thus, the existence of distinct POA "cold-detectors" is still undecided. This is not to say that responses to cooling cannot be obtained from the POA, since inhibition of the tonic firing of "warm-detectors" could serve this function.

A somewhat different line of research recently undertaken by electrophysiologists studying thermoregulation received its impetus from the observations of Feldberg and Myers (1964) that the biogenic amines, serotonin (5-HT) and norepinephrine (N.E.) injected into the cerebral ventricles or directly into the anterior hypo-thalamus produced marked changes in the rectal temperature of cats. The effects of amines on POA unit activity was first studied by Cunningham, et al. (1967) who found that epinephrine injected into the third ventricle of dogs depressed the firing of POA thermosensitive neurons, while 5-HT depressed the firing of both thermosensitive and insensitive cells.

More recently, Beckman and I (1970) have studied amine sensitivity of neurons using multibarreled micropipettes to apply the drugs electrophoretically directly onto the POA neurons recorded. Since it was reported that injected amines have opposite effects on rectal temperature in cats and rats (Feldberg and Lotti,1967; Beckman,1970),both species were studied. Some of the responses seen are illustrat-ed in figs. 5-8.

Fig. 5 shows the thermal response and drug sensitivity of a warm-sensitive cell of the interneuron-type which was transiently activated by acetylcholine (100 nano-amps current), and depressed by norepinephrine (100 nA current). Current passed from a barrel filled with NaCl had no effect, showing the specificity of the drug actions.

Fig. 6 shows the N.E. depression of firing of a high Q_{10}, thermodetector cell in a cat. The upper points (a) plot thermosensitivity in a control period; the lower points (b) show the thermal response obtained during continuous electrophoretic

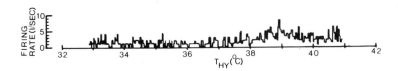

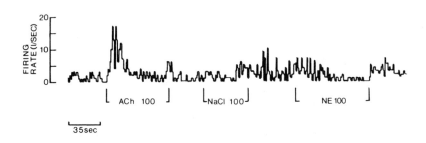

Fig. 5.   Amine sensitivity of a warm-sensitive interneuron in rat.
Responses to acetylcholine (ACh), current (NaCl) and nor-epinephrine (NE) are shown. 100 nA current used in each case. (From Beckman and Eisenman, 1970)

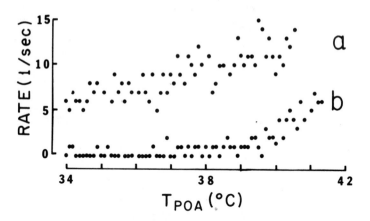

Fig. 6.   Effect of norepinephrine on thermal response of a high Q_{10}
          POA unit, in cat. Each point represents firing rate at 0.1
          C intervals. a: control. b: during N.E. electrophoresis.

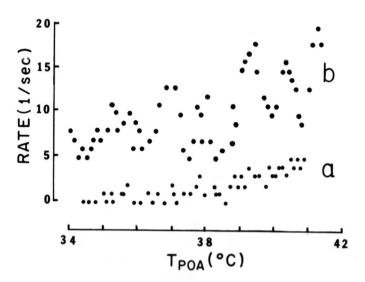

Fig. 7.   Glutamate activation of a POA neuron, in cat. a: control.
          b: during glutamate electrophoresis.

application of N.E. Firing was effectively zero until temperatures above 37-38 C
were reached. Note that during N.E. application the response looks like that of an
interneuron (non-linear or discontinuous) complicating the classification attempt.
Thermodetector-type cells showing amine sensitivity were,in fact, quite rare; only
2 out of 25 high Q_{10} cells tested were affected by applied amines. In both cases,
N.E. depression was seen.

Another substance that was tested in some instances was glutamate, which has
been shown to be excitatory to neurons in a number of different CNS structures.
More than half of the POA neurons tested were excited by glutamate. Fig. 7 shows
the response of a non-linear, warm-sensitive cell. Note that the glutamate and
thermal excitatory inputs appear to be additive. During glutamate application, the
thermal response appears continuous. The very marked variability of firing during
glutamate stimulation is characteristic of the excitation produced by this drug.
The cell appears to be over-driven and tends to fire in bursts.

As indicated above (cf. fig. 5), an artifact that must be controlled for when
using the micro-electrophoretic technique is the possible non-specific action of
current, per se, or of changes in the cell's ionic environment caused by the flow
of current, as opposed to specific drug effects. A barrel filled with NaCl was
used for such a control. Some cells did respond in a non-specific way to current
from all barrels, including the NaCl barrel, as shown in fig. 8. Such responses
were tabulated separately and not included as responses to drugs.

The changes in firing rate seen in this study are summarized in tables 2 and
3 for cats and rats respectively. For each neuron type, the number of cells that
increased (↑), decreased (↓), or showed no change in firing rate (0) with drug
application, current (±POLAR.) or glutamate (GLUT.) is tabulated. N is the total
number of cells of each type tested. Note that not all cells were tested with all

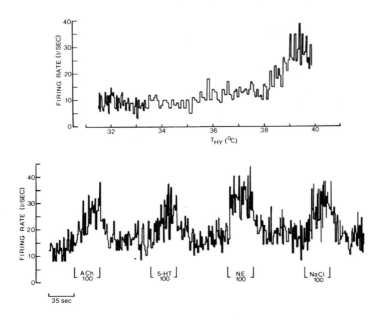

Fig.8. Non-specific, current effect on firing of a high Q_{10} POA
        neuron, in rat. Acetylcholine (ACh), serotonin (5-HT),
        norepinephrine (NE) and current (NaCl) all activate the
        cell. 100 nA current in each case. (From Beckman and
        Eisenman, 1970)

NEURON TYPE	N	N.E. ↑ ↓ 0			5-HT. ↑ ↓ 0			ACH. ↑ ↓ 0			+POLAR. ↑ ↓ 0			-POLAR. ↑ ↓ 0			GLUT. ↑ 0	
Q_{10}^1	47	0	7	39	2	4	36	1	1	44	1	4	25	4	1	31	5	2
Q_{10}^2	4	0	0	4	0	0	4	0	0	4	0	1	3	1	0	3	1	-
$Q_{10}^{>2}$	25	0	2	23	0	0	23	0	0	23	0	14	7	11	2	9	0	4
WARM INTER-NEURON	17	0	10	7	0	1	16	1	1	13	0	3	10	2	0	12	4	2
COOL INTER-NEURON	9	4	1	4	0	1	7	0	1	8	0	0	4	0	0	3	2	1
TOTALS	102	4	20	77 / 101	2	6	86 / 94	2	3	92 / 97	1	22	49 / 72	18	3	58 / 79	12	9 / 21

Table 2. Effects of monoamines and current on Cat's septal and POA units. The number of units showing a given response is indicated. N; total number of units of each type. ↑; increased firing rate. ↓; decreased firing rate. 0; no change.

NEURON TYPE	N	N.E. ↑ ↓ 0			5-HT. ↑ ↓ 0			ACH. ↑ ↓ 0			+POLAR. ↑ ↓ 0			-POLAR. ↑ ↓ 0		
Q_{10}^1 Q_{10}^2	23	0	3	14	0	5	15	0	0	4	0	4	9	2	2	6
$Q_{10}^{>2}$	7	0	2	5	0	1	5	0	0	6	3	3	1	1	2	0
WARM INTER-NEURON	4	0	2	2	1	1	2	0	0	0	0	1	3	0	0	0
COOL INTER-NEURON	0	-	-	-	-	-	-	-	-	-	-	-	-	-	-	-
TOTALS	34	0	7	21 / 28	1	7	22 / 30	0	0	10 / 10	3	8	13 / 24	3	4	6 / 13

Table 3. Effects of monoamines and current on Rat's septal and POA units. (See table 2, for symbols)

substances. In addition, the table does not specifically distinguish cells that responded to more than one test agent.

With regard to the thermosensitive cells, N.E. appears to inhibit warm-sensitive interneurons and excite cool-sensitive ones. The effect of 5-HT seems to be generally inhibitory, although surprisingly few cells responded. Only 2 out of 25 thermodetector-type cells were affected by the drugs (cf. fig. 6). Cells of this type were also not excited by glutamate. On the other hand, many high Q_{10}, linear cells were effected by polarizing current, cationic current depressing and anionic current accelerating the cell's firing rate. The lack of amine sensitivity and the responsiveness to polarization support the hypothesis that these detector-type cells do not have a synaptic input and are inherently active, as discussed earlier.

Whether the effects produced by current flow from the NaCl barrel are due to current (i.e., charge transfer across the membrane) or to some other ionic effect, such as an increased osmotic concentration in the cell's environment, as suggested by I.S. Edelman (personal communication) is not clear. If an osmotic effect were operating, I think that we would not see a reversal of response when the current was reversed, even taking account of the higher membrane permeability to Cl^-. The generally accepted assumption is that non-specific charge transfer is affecting the cell membrane potential. In any case, a non-specific action is operating, as opposed to a specific drug response.

The rat data are still rather sparse, but seem to be generally similar to those of the cat.

The results obtained with this technique do not correlate with the observations of Feldberg and Myers (1964) from microinjection studies. N.E. inhibition of warm-sensitive cells and activation of cool-sensitive ones would have the effect of raising central temperature, inhibiting heat loss outflow and exciting heat production. Microinjection of N.E. into a cat is reported to cause a fall in rectal temperature. Further, the reported species differences in response to injected amines has not, as yet, been seen in the unit recording studies. In this connection, Bligh, et al. (1971) have suggested that the observed species differences in response to injected amines may be due to ambient temperature effects rather than some inherent difference in synaptic organization. Clearly, further study is needed to resolve these contradictory observations.

At the beginning of this discussion, I suggested that much of the work in single unit recording served mainly to confirm or amplify information already developed by other techniques. There is, however, one observation made during microelectrode studies which was unexpected and, in fact, was contrary to the generally accepted scheme. In studying the effect of bacterial pyrogen on POA units, both Cabanac, et al. (1968), using rabbits, and I (Eisenman, 1969), using cats, observed that the slopes of the thermal response curves obtained from warm-sensitive units decreased during the pyrogen action. The implication of this finding was that during pyrogen induced fever POA thermosensitivity was decreased, probably by a direct action of the pyrogen on POA thermosensitive neurons. The usual description of thermoregulation during fever suggests that the regulator is set to a higher level with no change in regulatory capability.

In order to compare the results obtained in the unit studies with the thermoregulatory behavior of an intact animal, I have recently been studying metabolic responses of rabbits, chronically implanted with thermodes, to preoptic heating and cooling. This work was done in Dr. Hardy's laboratory at the Pierce Foundation. The rabbits were placed in a temperature controlled chamber at about 20 C. Oxygen consumption was determined by the open-circuit method using a Beckman Oxygen Analyzer to measure the oxygen concentration in the air stream flowing through a hood over the animal's head. Periods of POA heating and cooling were alternated to keep the core (rectal) temperature within 0.5C of the resting level (usually 38.5-39.3C). The control points on fig. 9 (x) indicate data from two experiments on the same rabbit, done on different days. The slope of the fitted line, -0.131 W·Kg^{-1}·$^{\circ}C^{-1}$, is taken as a measure of POA thermosensitivity.

For pyrogen runs, a bacterial pyrogen was injected into the lateral cerebral ventricle via a chronically implanted cannula. 60-90 minutes were allowed for the pyrogenic action to develop. During this time, the POA temperature was raised slowly as metabolic rate increased, so that rectal temperature did not rise; i.e., a fever was not allowed to develop. The metabolic response to POA temperature changes

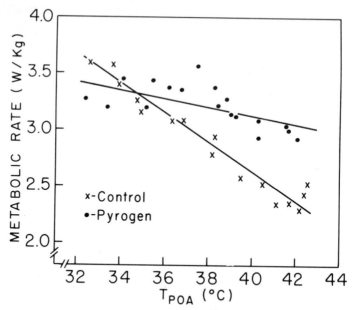

Fig.9. Metabolic response (watts/kg) of an unanesthetized rabbit
        to POA heating and cooling in 2 control experiments (x),
        and following intraventricular injection of bacterial
        pyrogen in 2 experiments (●).

was then studied. POA heating and cooling periods were again alternated to keep
rectal temperature from rising. Thus, inputs from non-controlled core sites, such
as the spinal cord, were the same for control and pyrogen experiments. The results
obtained from two such experiments on different days are shown in the figure (o),
the same rabbit having been used for control and pyrogen experiments. The slope of
the response curve during pyrogen action, -0.039 $W.Kg^{-1}.oC^{-1}$, is clearly less than
that during control runs. This result is taken to confirm the observations made
during the single unit studies.

     To conclude this discussion, I would like to make some suggestions for direc-
tions that might be taken in future research efforts.

     The first, I feel, may help resolve the problem of functional classification of
neurons. Unit activity has been studied in terms of average firing rate over some
short interval of time, usually on the order of seconds. There is reason to believe
that a closer look at the pattern of firing or its variability may provide further
insights into the functional role of a given cell.

     For some time, I have been looking at the inter-spike interval distributions of
POA neurons of various types and at the effect of temperature on these distributions. At fixed temperature, the time interval, in msec, between adjacent spikes in
a train is measured by computer and a tabulation made of the number of times a particular interval occurs. In such a tabulation the temporal sequence of the intervals is ignored and only the number of occurrences is recorded. The display of
such a tabulation is an inter-spike interval histogram, examples of which are shown
in fig. 10. Scales for all histograms in the figure are given on the last histogram.

     The three histograms shown in (A) were obtained from a posterior hypothalamic
unit with a discontinuous (non-linear) thermal response to posterior heating and
cooling; a presumed interneuron. The Gaussian distribution illustrated is quite
rare and is not characteristic of interneurons in general. Raising the local
temperature from 33.2 C to 37.2 C decreased the mean interval (increased the firing

rate) without changing the basic Gaussian shape of the distribution. The mean interval and standard deviation (S.D.) decreased proportionately. Heating to 40.9 C
caused no further change. (B) shows two histograms from a slightly thermosensitive
($Q_{10}2$) POA unit and (C) the distributions obtained from a highly sensitive, detector-type POA cell. The skewed distributions, with a pseudo-exponential decay are
characteristic of most of the units studied, regardless of functional classification. Increasing temperature produced a decrease in mean interval and a proportion-
ate decrease in S.D. with no change in the distribution shape, except in the last
histogram, C; 41.5 C. At this temperature, the distribution became bimodal, showing
two peaks. This is an indication of the "bursty" character of the neuron's discharge at this temperature. Although the cell acts as though it were being "over-
driven", its mean firing rate is only 27.6 spikes/sec. Within a burst, much higher
instantaneous frequencies are reached.

  Another piece of information that can be derived from such analysis, in addition
to distribution shapes, relates to the variability of the unit's firing. If we
propose that different processes are responsible for the ongoing activity in detector cells as compared to interneurons, then perhaps this will be reflected by

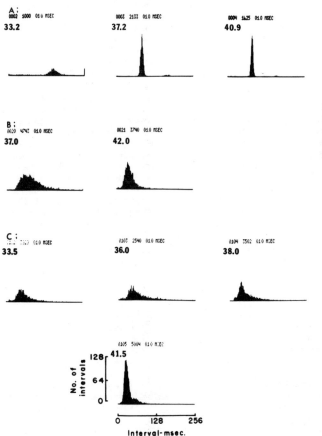

Fig. 10. Inter-spike
interval histograms
for three hypothalamic
neurons at various
temperatures.

differences in the discharge variability in spike trains generated by the different neuron types. This does not suggest that responses in heat-loss pathways are neuronally coded by a pattern different from that in heat-production and conserva-tion systems. The much simpler hypothesis, that different generator processes in neurons may also generate different activity patterns, was explored. The co-efficient of variability (C.V.=S.D./mean) was calculated for about 200 inter-spike interval histograms obtained from units of all functional types (insensitive, detector, interneuron) at various temperatures. Unfortunately, no significant differences were seen between the coefficients of these groups.

These preliminary studies, therefore, have not detected any characteristics re-lating either probability distribution functions or variability of firing to pre-sumed function for POA neurons. I feel, however, that a more sophisticated analysis of these data may reveal fundamental differences.

Another line of investigation which is technically feasible is the application of the single unit recording method in unanesthetized, behaving animals. The ad-vantage of this preparation is that it will eliminate the depressant effect of anesthetic on the activity of brain-stem structures and will allow recording of effector, regulatory responses as well as unit activity. Steps in this direction have been taken by Hellon (1970) who has recorded unit activity in unanesthetized rabbits. Guieu and Hardy (1970) have also found that they could detect some effector responses (vasodilation, polypnea) during unit recording and POA heating in lightly anesthetized rabbits. When the effector output of the intact regulatory system can be recorded and correlated with activity of the neurons driving this output, we will have the best possible preparation for unraveling the neural net-works controlling body temperature. The techniques are available; the necessary background of information and experience has been accumulated. It remains only to apply them.

One final suggestion of an area that deserves more attention is the motor con-trol of thermoregulatory effector organs. There is, currently, an increased inter-est in peripheral autonomic nervous system function. Study of the activation of autonomic neurons by thermal stimuli, while technically difficult, would provide useful information on the organization of the several regulatory functions served by these neurons. Similarly, the thermal input to the respiratory "centers" and the motor organization of shivering have not been completely analyzed.

In summary, much work has been done in studying the CNS neurons forming the control net for body temperature, using the single unit recording technique. The data derived from these studies have helped to clarify our understanding of the patterns of activity of thermosensitive neurons at various CNS levels, and of the levels and degrees of complexity of neuronal interactions. Using this information, it should now be possible to devise and to execute experiments to define more clearly the role of CNS and peripheral neurons in thermoregulation.

REFERENCES

Barker, J.L. and D.O. Carpenter, (1970), Thermosensitivity of neurons in the sensorimotor cortex of the cat, Science 169, 597.
Beckman, A.L., (1970), Effect of intrahypothalamic norepinephrine on thermo-regulatory responses in the rat, Am. J. Physiol. 218, 1596.
Beckman, A.L. and J.S. Eisenman, (1970), Microelectrophoresis of biogenic amines on hypothalamic thermosensitive cells, Science 170, 334.
Birzis, L. and A. Hemingway, (1957), Efferent brain discharge during shivering, J. Neurophysiol. 20, 156.
Bligh, J., W.H. Cottle and M. Maskrey, (1971), Influence of ambient temperature on the thermoregulatory responses to 5-hydroxytryptamine, noradrenaline and acety-choline injected into the lateral cerebral ventricles of sheep, goats and rabbits, J. Physiol. (Lond.) 212, 377.
Cabanac, M., J.A.J. Stolwijk and J.D. Hardy, (1968), Effect of temperature and pyrogens on single-unit activity in the rabbit's brain stem, J. Appl. Physiol. 24, 645.

Cunningham, D.J., J.A.J. Stolwijk, N. Murakami and J.D. Hardy, (1967), Responses
    of neurons in the preoptic area to temperature, serotonin and epinephrine,
    Am. J. Physiol. 213, 1570.
Edinger, H.M. and J.S. Eisenman, (1970), Thermosensitive neurons in tuberal and
    posterior hypothalamus of cats, Am. J. Physiol. 219, 1098.
Eisenman, J.S., (1969), Pyrogen-induced changes in the thermosensitivity of septal
    and preoptic neurons, Am. J. Physiol. 216, 330.
Eisenman, J.S. and D.C. Jackson, (1967), Thermal response patterns of septal and
    preoptic neurons in cats, Exper. Neurol. 19, 33.
Eisenman, J.S. and H.M. Edinger, (1971), Neuronal sensitivity, Science 172, 1360.
Feldberg, W. and R.D. Myers, (1964), Temperature changes produced by amines in-
    jected into the cerebral ventricles during anaesthesia, J. Physiol. (Lond.)
    175, 464.
Feldberg, W. and V.J. Lotti, (1967), Temperature responses to monoamine and an
    inhibitor of MOA injected into the cerebral ventricles of rats, Br. J.
    Pharmacol. Chemother. 31, 152.
Guieu, J.D. and J.D. Hardy, (1970), Effects of heating and cooling of the spinal
    cord on preoptic unit activity, J. Appl. Physiol. 29, 675.
Hardy, J.D., R.F. Hellon and K. Sutherland, (1964), Temperature-sensitive
    neurones in the dog's hypothalamus, J. Physiol. (Lond.) 175, 242.
Hellon, R.F., (1970), The stimulation of hypothalamic neurones by changes in
    ambient temperature, Pflügers Arch. 321, 56.
Magoun, H.W., F. Harrison, J.R. Brobeck and S.W. Ranson, (1938), Activation of
    heat loss mechanisms by local heating of the brain, J. Neurophysiol. 1, 101.
Nakayama, T., J.S. Eisenman and J.D. Hardy, (1961), Single unit activity of
    anterior hypothalamus during local heating, Science 134, 560.
Nakayama, T., H.T. Hammel, J.D. Hardy and J.S. Eisenman, (1963), Thermal stimula-
    tion of electrical activity of single units of the preoptic region, Am. J.
    Physiol. 204, 1122.
Nakayama, T. and J.D. Hardy, (1969), Unit responses in the rabbit's brain stem to
    changes in brain and cutaneous temperature, J. Appl. Physiol. 27, 848.
Nakayama, T. and T. Hori, (1971), Effects of anesthetics and Pyrogen on thermally
    sensitive neurons in the brainstem, Int. J. Biometeorol., in press.
Nutik, S.L., (1971), Effect of temperature change of the preoptic region and skin
    on posterior hypothalamic neurons, J. Physiol. (Paris) 63, 368.
Parmeggiani, P.L., C. Franzini and P. Lenzi, (1971), Sleep related changes in
    hypothalamic unit activity and specific thermal responses, Proc. Int. Union
    Physiol. Sci. (XXV Int. Cong.) 9, 440.
Wit, A., and S.C. Wang, (1968), Temperature-sensitive neurons in preoptic/anterior
    hypothalamic region: effects of increasing ambient temperature, Am. J. Physiol.
    215, 1151.
Wünnenberg, W. and J.D. Hardy, (1971), Responses of units in the posterior hypo-
    thalamus to local temperature changes in the preoptic region, the posterior
    hypothalamus and the spinal cord, Int. J. Biometeorol., in press.

CENTRAL TRANSMITTERS AND THERMOREGULATION

RICHARD HELLON

National Institute for Medical Research

London NW7 1AA

U.K.

In this essay, the aim will be to assess what is known about neurotransmitter substances in the temperature controller and try to decide how far this knowledge increases our understanding of the working of that controller.

The whole subject of neurotransmitters dates from the early years of this century when Elliot pointed out the close similarity between the effects of sympathetic nerve stimulation and those of administered adrenaline. He suggested that the nerve endings actually released an adrenaline-like substance as an intermediary of their action. Less than twenty years later Loewi showed that a chemical intermediary was also responsible for the slowing action of the vagus on the frog's heart and this was subsequently shown to be acetylcholine. Later the same substance was shown to be the transmitter at the neuromuscular junction and in sympathetic ganglia by Dale, Feldberg, Gaddum and their colleagues. The application of the concept of chemical transmission from these peripheral sites to the synapses of the central nervous system began after World War II, but so far there is firm evidence identifying the transmitter substance only at very few synapses. Great care has to be taken in concluding that the action of a substance on a post-synaptic site is a true physiological one and not simply a pharmacological curiosity. Various criteria should be satisfied before a substance can be seriously regarded as a transmitter of synaptic action and these have been listed by McLennan (1963) as follows:

(1)   The substance must occur in those neurons whose action it transmits at the synapses it forms with ensuing neurons.

(2)   The neuron must possess an enzymatic mechanism for the synthesis of the substance.

(3)   There must be a system present for the inactivation of the substance either by enzymatic action or by re-uptake into the presynaptic terminals.

(4)   Application of the substance to the postsynaptic structure must mimic the action of stimulation of the neuron.

(5)   During stimulation the substance must be detectable in extracellular fluid collected from the region of the activated synapses.

(6)   Pharmacological agents which interfere with the transmission at a synapse must similarly affect the action of the substance when applied artificially.

In the few sites where transmitters have been almost certainly
identified in the CNS, the synapses have been defined anatomically
and the input and output pathways have been accessible for stimulation and recording.   For example, this is the case for the
Renshaw cells in the spinal cord where acetylcholine is almost
certainly an excitatory transmitter from the endings of the collaterals from the motor axons (Eccles, Fatt & Koketsu, 1954).
Inhibitory transmitters have proved to be elusive in the CNS but
there is recent evidence that noradrenaline is most likely an
inhibiting substance on the Purkinje cells of the cerebellar cortex
(Hoffer, Siggins & Bloom, 1971).   In the case of the temperature-
regulating neurons in the hypothalamus, the pathways are virtually
unknown and none of the synapses have been identified.   Thus at
first sight it might appear hopeless to try and apply McLennan's six
criteria to such a complex web of neurons.   However there are two
factors which ease our task of looking for possible transmitters in
this sytem.   The first of these is the rare stroke of anatomical
fortune that the neurons concerned lie close to the walls of the
third ventricle and hence can be reached through diffusion by drugs
introduced into that space.   Secondly the temperature controller has
well defined actions on body temperature and the mechanisms for
controlling it, so that the output of the controller can be judged
by relatively simple means and in the absence of anaesthetics.

THE MONOAMINES

Since 1954 it has been known that in the hypothalamus there
are particularly high concentrations of the monoamines, adrenaline,
noradrenaline (Vogt, 1954) and 5-hydroxytryptamine (5-HT)(Amin et
al, 1954).   By 1961 von Euler was able to suggest: '....... that
these putative central neurohumors .... might exert an influence
on the setting mechanism of the body thermostat'.   The first
evidence that this might be so came from the experiments of Feldberg
and Myers (1964, 1965).   They injected microgram doses of the
catecholamines and 5-HT into the cerebral ventricles or the anterior
hypothalamus of cats.   The effects of the two catecholamines were
similar and resulted in a fall of rectal temperature due to skin
vasodilation and reduced heat production.   The opposite result was
caused by 5-HT with a rise of temperature caused by shivering, vasoconstriction and piloerection.   Fig. 1 shows the results from one of
their experiments.   There is a brief change in rectal temperature
with the catecholamines, but the long term change with 5-HT is
unusual and has not yet been explained satisfactorily.   These were
astonishing findings when one considers the probable complexity of
the neuronal network into which the substances were introduced.
Feldberg and Myers put forward the hypothesis that temperature
regulation is achieved by a balance of these opposing actions due to
the relative rates of release of catecholamines and 5-HT in the
hypothalamus.

Although it was implied that these substances were acting as
neuro-transmitters, the results provide no evidence that this was
the case.   The likelihood that they could be transmitters is made
stronger by the fluorescent microscopic technique which shows that
in the hypothalamus these monoamines are mainly localized in the
terminal nerve endings, which is where a transmitter would be
expected (Carlsson, Falck & Hillarp, 1962; Dahlström & Fuxe, 1964;
Andén, Dahlström, Fuxe & Larsson, 1965).

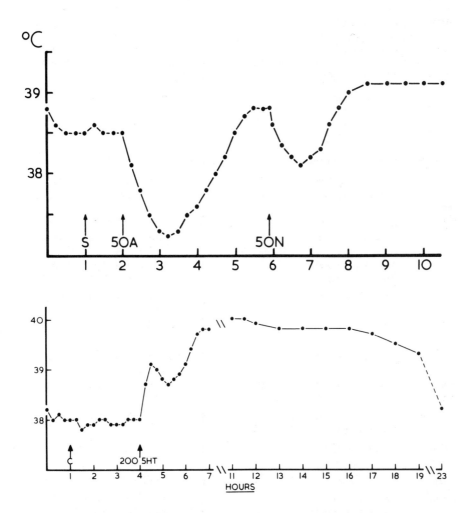

Fig. 1. Changes in rectal temperature in two unanaesthetized cats.
Each arrow indicates an injection of 0.1 ml. into a lateral
ventricle. In the upper record, the successive injections contained 0.9% NaCl, 50 µg adrenaline and 50 µg noradrenaline; in
the lower record, the injections were of 100 µg creatinine
sulphate and 200 µg 5-HT. From Feldberg & Myers (1964).

        If the monoamines were acting as transmitters, the enzymes for
their synthesis and destruction should also be present in the
hypothalamus and there is evidence that this is true (see Feldberg,
1968). We shall consider the effects on temperature of interfering
with these enzymes in another section. The concept of Feldberg and
Myers has stimulated a great deal of experimentation which is still
continuing at the present time.

SPECIES DIFFERENCES

The results on cats were confirmed on dogs (Feldberg, Hellon & Myers, 1966) and on rhesus monkeys (Feldberg, Hellon & Lotti, 1967). In all three species the catecholamines activated heat loss and so lowered body temperature, while 5-HT had the opposite action.

Different responses were found when these substances were tested in other species. In rabbits, there was a rise in temperature

Species	Route of administration	T_b Change NA	5-HT	Authors
Cat	Lateral ventricle	↓	↑	Feldberg & Myers (1963, 1964)
	Lateral ventricle	↓	↑	Ruckebusch, Grivel & Laplace (1965)
	Lateral ventricle		↓	Kulkarni (1967)
	Lateral ventricle		↑ ↓	Banerjee, Burks & Feldberg (1968)
	Ant. hypothalamus	↓	↑	Feldberg & Myers (1965)
	Preoptic area		↓	Jacobson (1967)
Dog	Lateral ventricle	$\downarrow_1$	$\uparrow_1$	Feldberg, Hellon & Myers (1966)
	Third ventricle	↓	↑	Feldberg, Hellon & Lotti (1967)
Rabbit	Lateral ventricle	↑	↑ ↓	Ruckebusch, Grivel & Laplace (1965)
	Lateral ventricle	↑	↓	Cooper, Cranston & Honour (1965)
Sheep	Lateral ventricle	↑	↑ ↓	Ruckebush et al. (1965)
	Lateral ventricle	↑	↓	Bligh (1966)
Goat	Third ventricle		↓	Andersson, Jobin & Olsson (1966)
Ox	Lateral ventricle	$\downarrow_4$	↓	Findlay & Thompson (1968)
Rat	Lateral ventricle	$\uparrow_2\downarrow_3$	↓	Feldberg & Lotti (1967)
	Lateral ventricle	↑	↓	Myers & Yaksh (1968)
Mouse	Lateral ventricle	↓	↓	Brittain & Handley (1967)
Monkey	Lateral ventricle	↓	$\uparrow_2\downarrow_3$	Myers & Sharpe (1967)
	Ant. hypothalamus	↓		Myers & Yaksh (1969)
	Third ventricle	$\downarrow_1$	$\uparrow_1$	Feldberg et al. (1967)

Fig. 2. A summary of the effects on body temperature of intra-ventricular or intrahypothalamic injections of noradrenaline and 5-HT in different species. The arrows indicate the direction of change in body temperature; a light arrow shows a feeble or irregular response. Subscripts: 1 = anaesthetized, 2 = low dose, 3 = high dose, 4 = at low or moderate ambient temperatures only. From Bligh, Cottle & Maskrey (1971), which also gives the references not quoted in the present paper.

with intraventricular noradrenaline and a slight fall with 5-HT
(Cooper, Cranston & Honour, 1965). Other species showed different
responses again: the mouse showed hypothermia with both of the
monoamines. A summary of the results obtained up to now has been
made by Bligh, Cottle and Maskrey (1971) and is shown in Fig. 2.
Even in the same species, there are conflicting reports. For example
injections of 5-HT in cats have also been found to cause a marked
fall in core temperature in contrast to the results in Fig. 1.

      Part of the explanation for this great diversity may be genuine
species differences in the transmitters used for a particular
function. But on the other hand, in some of these experiments only
one, usually a large, dose of the amine was given. This could have
caused synaptic blockade rather than excitation. Obviously it is
necessary to establish a proper dose-response relationship.

      Another important factor in determining an animal's responses
to injected monoamines is the prevailing ambient temperature. If
a particular pathway, say that concerned with heat production and
conservation, is already active because of cool surroundings, then
a drug which at neutral temperatures causes hyperthermia would be
expected to have less effect or none at all. Such an interaction
between ambient temperature and intraventricular injections was first
demonstrated by Findlay and Thompson (1968). Using oxen, they found
that noradrenaline, which is hypothermic in this animal, had no
action at 30°C when the heat loss pathways were already activated,
but at -1°C when these pathways were clearly not functioning, the
noradrenaline did lower body temperature and lowered the plasma
levels of free fatty acid.

      Bligh and his colleagues have made a systematic study of the
effects of ambient temperature using sheep, rabbits and goats.
They found that intraventricular 5-HT given at low temperatures
increased heat loss by panting and also decreased heat production by
inhibiting shivering. At high temperatures where the animals were
already panting, 5-HT had no effect. This indicates that 5-HT has
an excitatory action on the heat loss pathway with an inhibitory
influence on heat production. Noradrenaline did not behave in this
specific way but had a general inhibitory action on whatever thermoregulatory activity was present. The panting at high ambient
temperatures was depressed by noradrenaline so that body temperature

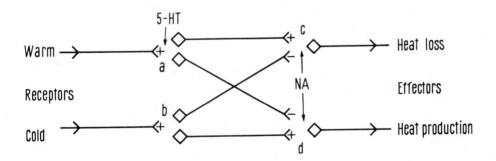

Fig. 3. A neuronal model based on the effects of intraventricular
    injections of 5-HT and noradrenaline in sheep under warm and cold
    conditions. From Bligh & Cottle (1969).

rose, whereas shivering in low ambient temperatures was reduced by
noradrenaline and body temperature fell. This inhibitory property
of noradrenaline is in keeping with its action on other synapses
such as the mitral cells in the olfactory bulb, the Purkinje cells
in the cerebellum or in the hypothalamus itself. The results
obtained with 5-HT and noradrenaline at different ambient temperatures form the basis for a simple neuronal model suggested by Bligh
and Cottle (1969). It is shown in Fig. 3. The question of what
substance is the possible transmitter at synapse 'b' will be discussed in a later section. It is not yet certain how far this model
can be extended to species other than sheep, rabbits and goats. At
the moment it seems unlikely to apply to the cat-dog-monkey group
because on cats, Feldberg and I (unpublished data) have been unable
to find a reversal of the action of noradrenaline by changing the
ambient temperature. The large fall of body temperature produced
in the cold by intraventricular noradrenaline was not converted to
a rise in the heat when heat-loss effectors were already mobilized.
There was just a slight fall.

       Taken together, all the results we have discussed so far
indicate that the monoamines, noradrenaline and 5-HT, could act as
central transmitters in the control of temperature regulation, and
that there seem to be real differences in the way they act in
different species.

ENDOGENOUS MONOAMINES

       Even the doses of the monoamines which have to be given intra-
ventricularly are large compared with the amounts present in the
hypothalamus. Although it can be argued that only a small fraction
of the dose injected actually reaches the postsynaptic membrane, it
seems necessary to test whether the responses to exogenous amines
are the same as those evoked by the animal's own endogenous supplies.
This can be done by depleting the presynaptic stores, by interfering
with the mechanisms which inactivate the released transmitter, or by
blocking transmitter action on postsynaptic membranes. There is not
space to mention all the work which has been done using these techniques, but the balance of the evidence confirms the original hypothesis of Feldberg and Myers, after allowance is made for the species
differences. I would like to give just one example of some recent
work concerning noradrenaline (Cranston, Hellon, Luff & Rawlins,
1971). This transmitter is mainly inactivated by re-uptake into the
presynaptic endings and this process can be inhibited by the anti-
depressant drugs, imipramine and N-des-methylimipramine. If
noradrenaline were a transmitter in the temperature regulation
pathways, an intraventricular injection of these drugs should allow
released noradrenaline to accumulate at synapses and so mimic the
action of exogenous noradrenaline. Furthermore the effects in cats
and rabbits should be opposite in direction. The results which are
shown in Fig. 4 confirmed these predictions: body temperature fell
in cats and rose in rabbits and in addition these temperature changes
were much diminished if the endogenous stores of noradrenaline had
previously been depleted by alpha-methyl paratyrosine. These
experiments demonstrate that an animal's own noradrenaline causes
temperature effects which are the same as those produced by intraventricular injections of the amine and this increases the likelihood that it may be acting as a transmitter.

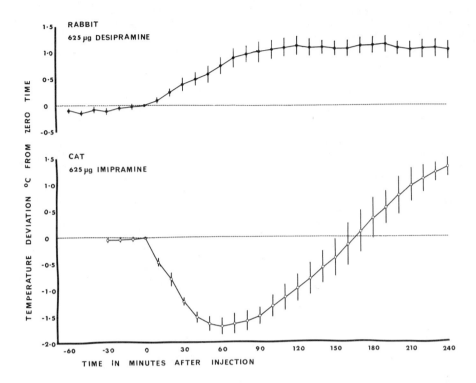

Fig. 4.   Mean changes in rectal temperature of five rabbits and four cats following injections into a lateral ventricle of desipramine or imipramine at time zero.   Vertical markings indicate $\pm$ 1 S.E. of mean.   From Cranston, Hellon, Luff & Rawlins (1971).

ACETYLCHOLINE

Since acetylcholine (ACh) is one of the chief transmitter substances at peripheral synapses, it is natural that there should have been many attempts to find if it is also a central transmitter. Several tests with ACh alone given intraventricularly failed to show an effect on body temperature and this was probably due to the rapid destruction of ACh, before any temperature action was seen.   However when the ACh was protected with eserine or when cholinomimetics like carbachol were used, then it became evident that cholinergic synapses were probably concerned with temperature regulation.   Once again we have species differences, as was the case with 5-HT and noradrenaline.   Rats and mice form one group and give a hypothermic response to cholinomimetics (Kirkpatrick & Lomax, 1969; Hulst & de Wied, 1967; Myers & Yaksh, 1968).   Another group is formed by sheep, goats and monkeys which respond with a rise in temperature, while rabbits do not belong to either group and show no response.

Bligh, Cottle and Maskrey, (1971), gave intraventricular injections of ACh, eserine or carbachol at various ambient temperatures and they found similar responses in both sheep and goats.

The action was what would be expected for an excitatory transmitter
on the heat-production pathway. Under cool conditions shivering was
increased and under warm conditions respiratory heat loss was
reduced, both of these actions leading to a rise in body temperature.
We can now recall Bligh's model shown in Fig. 3 and insert the
missing substance at synapse 'b' which is apparently acetylcholine.

MICROINJECTION STUDIES

        The intraventricular injections we have been considering so
far give little indication as to where the injected substance may
act in the brain. By giving microinjections into the substance of
the anterior hypothalamus Cooper, Cranston and Honour (1965) found
the same temperature effects in rabbits from noradrenaline and 5-HT
as when these substances were given intraventricularly. Similar
results were seen in cats (Feldberg & Myers, 1965). Thus at least
part of the action of the monoamines is on neurons in the anterior
hypothalamus. Myers and Yaksh (1969) have made a more systematic
survey of the hypothalamus of monkeys. The substances they used in
these experiments on conscious monkeys were not only noradrenaline
and 5-HT, but also ACh. The exact point at which each of these
three substances acted to affect body temperature was found by
making bilateral injections in volumes of 1μl or less. Nearly all
the sites for the hyperthermic effect of 5-HT and the hypothermic
effect of noradrenaline were closely grouped together in the anterior

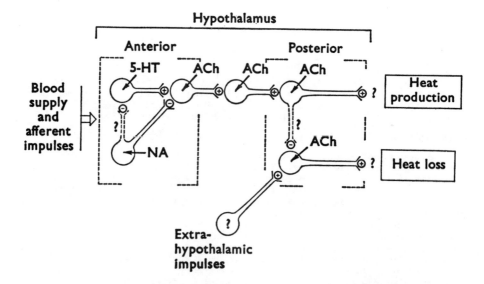

Fig. 5. A neuronal model based on the effects on body temperature
    of microinjections of 5-HT, noradrenaline or ACh into various
    hypothalamic regions in monkeys. 5-HT cells respond to cooling
    and activate heat production through a cholinergic pathway to the
    posterior hypothalamus. Noradrenaline cells respond to warming
    and inhibit the heat production pathway. This inhibition then
    allows the cholinergic heat loss pathway in the posterior
    hypothalamus to become active, possibly by stimulation from
    extra-hypothalamic sites. From Myers & Yaksh (1969).

part of the hypothalamus lying between the anterior commissure and
the optic chiasma. In contrast to this, the points where ACh was
effective were fairly uniformly scattered throughout the hypothalamus. Usually this drug caused a rise of body temperature but one
or two sites in the posterior hypothalamus were found where it
caused a fall. From these results another chemical model was
produced and this is shown in Fig. 5. When heat production is called
for, the cells responding to 5-HT are activated and the impulses
passed down an acetylcholine chain to activate mechanisms such as
shivering. On the other hand the heat loss pathway is not so
clearly defined and chiefly seems to consist of inhibition by
noradrenaline on cells of the heat production pathway. How far
these results can be applied to other species is uncertain at the
moment, but it would certainly be very worthwhile to try similar
experiments in other animals.

PROSTAGLANDINS

        These substances occur quite widely in the CNS, including the
hypothalamus, and in 1970 Milton and Wendlandt tried the effects on
temperature of injecting various prostaglandins into the third
ventricle of unanaesthetized cats. They found that only one of the
prostaglandins, E_1, was effective in changing body temperature and
this caused a steep rise with doses as small as 20 or 10 nanograms.
This is a very small dose compared with the other experiments we
have been talking about earlier. These results were confirmed by
Feldberg and Saxena (1971) who also did experiments on rabbits and
rats. In all three species there was a rapid rise in body temperature due to extremely vigorous shivering after very small doses of
prostaglandin were given. It is too early to say whether this
prostaglandin has any physiological role in temperature regulation
but, as we shall see in a moment, some substance like a prostaglandin can be detected when the ventricles of an animal are perfused. However the fact that cats, rats and rabbits all respond in
the same way is quite a hopeful sign.

PERFUSION STUDIES

        To establish with a higher probability that a substance may be
playing the role of a neurotransmitter in temperature regulation, it
is necessary to do more than show that exogenous or even endogenous
application can drive the body temperature up or down. The candi-
date transmitter must also be shown to be released when the hypothalamic synapses might be expected to be activated. For example,
a monkey, which has a hyperthermic response to 5-HT, might be
expected to show a release of 5-HT when exposed to the cold. If
this could be done it would considerably strengthen the argument
that 5-HT is acting as a neurotransmitter in this system. The
problem has been looked at with two techniques: by perfusion of the
third ventricle or by using Gaddum's method of direct push-pull
perfusion of the actual hypothalamic tissue.

        The first attempts at this sort of experiment were made by
Feldberg and Myers (1966) who found 5-HT in the perfusate from the
third ventricle of anaesthetized cats. Sometimes the output of 5-HT
increased as the anaesthesia lightened and the cat's temperature
began to rise. Another substance was also found in the perfusate
which behaved in the bioassay as if it was prostaglandin E_1.

        A novel method of bioassay for ventricular perfusion was
introduced by Myers (1967a). He collected the perfusate from the

third ventricle of one monkey, the donor monkey, and passed it into
the third ventricle of a second monkey, the recipient. When the
donor monkey was heated the temperature of the recipient monkey was
found to fall, with the opposite result if the donor monkey was
cooled (Myers, 1967b). However, this sort of result was not
consistently obtained and Myers and Sharpe (1968) devised another
system in which bilateral push-pull cannulae were implanted into the
anterior hypothalamus of a pair of monkeys, so that perfusate could
be collected from the nervous tissue of one monkey and directly
infused into corresponding sites in the second monkey. As in the
earlier experiments, heating of the donor monkey caused the
temperature of the recipient to fall abruptly, while cooling of the
donor caused the temperature of the recipient to rise quickly.

    Now these elegant results raise the question: what substances
are being transferred across from one monkey to another to cause
these effects? Bearing in mind the species differences, you would
expect that the perfusate from a cooled monkey would contain
increased amounts of 5-HT while that from the heated monkey should
contain more noradrenaline. Analysis of these substances is not yet
complete, but recently Myers and Beleslin (1971) have been able to
show that increased amounts of 5-HT are found in perfusate collected
from a cooled monkey. Heating a monkey had little or no effect on
the output of 5-HT. Some of their results are shown in Fig. 6,
where the change in output of 5-HT in the perfusate is given along
with the anatomical positions of the perfusion cannulae. In several
of these positions, external cooling caused a large increase in the
concentration of released 5-HT. The region shown in Fig. 6 is just
the same area in which injections of a few µg of 5-HT cause a
monkey's temperature to rise. This is strong evidence that 5-HT may
be acting as a transmitter in the control of heat production in the
monkey. But there is a discrepancy in the amounts of 5-HT released
into the perfusion fluid, which are measured in nanograms, and the
amounts of 5-HT which need to be injected to cause a rise in
temperature, which are in micrograms. Possibly an additional
compound is released along with the 5-HT which would account for
the hyperthermia seen in the cross-perfusion experiments using two
monkeys. A prostaglandin would fit the requirements, but tests in
the bioassay of the perfusate seem to rule this out.

UNIT ACTIVITY

    Next we must consider a further refinement of technique in
which the activity of an individual neuron is tested first for its
temperature sensitivity and then for its sensitivity to the various
possible transmitters.

    The first experiments on these lines were made by Cunningham,
Stolwijk, Murakami and Hardy (1967). Using dogs they found that
intraventricular injections of both noradrenaline and 5-HT tended to
depress the activity of eight temperature-sensitive cells. This is
not the sort of result that would have been predicted on the evidence
we have considered up to this point. On that basis either noradrenaline or 5-HT would be expected to have a specific action on these
temperature cells, but this was not the case. It is possible there
may have been a sampling error and that if more than eight neurons
had been tested, some would have been found which were specifically
sensitive to the injections.

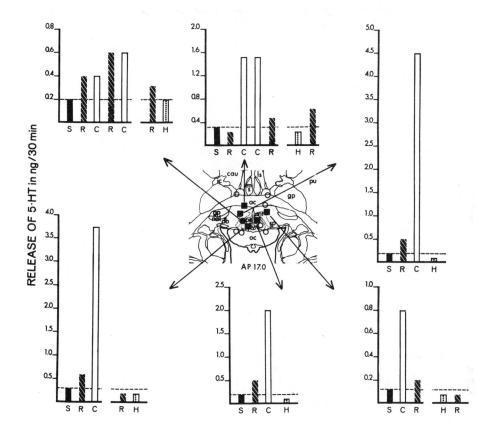

Fig. 6. Perfusion sites in the anterior hypothalamus and 5-HT assay
   results on rhesus monkeys. The central diagram shows a coronal
   section passing through the anterior commissure (ac) and the
   optic chiasma (oc); perfusion sites are shown where 5-HT output
   changed (■), remained unchanged (●), or was not detected (O).
   The surrounding assays show 5-HT (ng/30 min) released from these
   sites; S columns and horizontal broken lines indicate minimum
   sensitivity of isolated organ strip for each assay; R columns
   show control or resting release; C columns show release during
   external cooling; H columns show release during external heating.
   From Myers & Beleslin (1971).

        If we could apply the candidate transmitters directly on to the
synaptic regions, then the results should be easier to interpret.
This can be done by using the technique of microiontophoresis from a
multi-barrelled electrode and the first attempt to do this in the
hypothalamus was by Bloom, Oliver and Salmoiraghi (1963). They did
not use thermal testing, but they did find that many cells had their
spontaneous firing rate changed when ACh, 5-HT or noradrenaline were
administered. The techniques of microiontophoresis and thermal

testing were combined by Beckman and Eisenman (1970). They tested a
neuron first for its responses to changes in local brain temperature
and then for its responses to ejection of noradrenaline, 5-HT and
ACh. One type of warm-sensitive cell was highly temperature-
sensitive and responded equally well to the three drugs but also to
the iontophoresing current, which means that the 'drug responses'
were actually due to the current. The implication is that these
neurons may have no synaptic input and could be actual 'end organs'
for transducing temperature. The other group of temperature cells
were classed as interneurons by Beckman and Eisenman and they were
either warm-sensitive or cold-sensitive. They responded to the
drugs by altering their firing rate but not to current flow. There
was a correlation between direction of the thermal sensitivity and
the responses to the drugs. The warm-sensitive interneurons were
accelerated by ACh and slowed by noradrenaline. The firing rate of
the cold-sensitive interneurons was accelerated by noradrenaline.
The same results were obtained from rats and cats.

Now the responses of the rat's interneurons agree with what
would be predicted from microinjections into the anterior hypothalamus. ACh activates heat loss mechanisms and on single cells it
increases the firing rate of warm-sensitive cells. Noradrenaline, on
the other hand, appears to activate heat production mechanisms, at
least in small doses, and it has the predicted effect on neurons of
inhibiting warm-sensitive cells and exciting cold-sensitive ones.

Exactly the same responses at the single cell level were found
in cats, but in this animal noradrenaline injections into the ventricles have the opposite effect to that in rats. This means that
there is a discrepancy between the action of this amine when
recorded on single neurons and when given as a gross injection.
This is a question which has still to be resolved. Clearly many
more recordings will have to be made using this powerful technique
and possible difficulties due to anaesthetics will have to be ruled
out, perhaps by using the diencephalic island technique (Cross &
Kitay, 1967).

CONCLUSIONS

It now remains to look back over all this evidence and try to
decide what place it might have in that part of the nervous system
which is controlling body temperature.

Most of the evidence is highly circumstantial but at the same
time it forms an impressive and fairly coherent assembly of observations. We have seen that there are naturally-occurring substances
in nerve endings in that part of the hypothalamus which is known to
be involved in temperature control. When the concentration of one
of these substances is locally raised it gives rise to a coordinated
series of physiological and behavioural responses which are unlikely
to be a pharmacological accident. But real proof that we are dealing with what appear to be neurotransmitters comes from the cross-
perfusion and microperfusion experiments of Myers' laboratory, where
changes in 5-HT output can be caused by external heating and cooling.
With more sensitive assay methods it should be possible to do the
same with noradrenaline. Also much more detailed understanding will
come when more microiontophoresis experiments have been done.

But, granting that 5-HT, noradrenaline and acetylcholine are
some or all of the transmitters concerned, what does this tell us
about the basic functioning of the temperature controller? It seems

that the chief value of this evidence is twofold. First it serves as a microanatomical tool which shows where particular groups of synapses can be found. This would have been quite impossible with the usual anatomical methods. Secondly it has led to the 2 models we have seen and the testing of these, particularly at the single unit level, should lead to much valuable new information.

It must never be forgotten that we are dealing with a temperature control system and therefore, although knowledge of the chemical links between elements is valuable, it is temperature which has to be sensed and controlled at the appropriate level. Let us make a comparison with another biological control system – that which governs the amount of light entering the eye. Light falls on the retina, gives rise to a whole series of synaptic chemical events and finally ACh is released to cause the sphincter pupillae to contract until the optimum amount of light is admitted. Light is sensed and then controlled. In my view the same attitude should be taken of the temperature control system.

REFERENCES

Amin, A.N., Crawford, T.B.B. & Gaddum, J.H. (1954) The distribution of substance P and 5-hydroxytryptamine in the central nervous system of the dog. J. Physiol. 126, 598.

Andén, N.E., Dahlström, A., Fuxe, K. & Larsson, K. (1965) Mapping of catecholamine and 5-hydroxytryptamine neurons innervating the telencephalon and diencephalon. Life Sci. Oxford 4, 1275.

Beckman, A.L. & Eisenman, J.S. (1970) Microelectrophoresis of biogenic amines on hypothalamic thermosensitive cells. Science 170, 334.

Bligh, J. & Cottle, W.H. (1969) The influence of ambient temperature on thermoregulatory responses to intraventricularly injected monoamines in sheep, goats and rabbits. Experientia 25, 608.

Bligh, J., Cottle, W.H. & Maskrey, M. (1971) Influence of ambient temperature on the thermoregulatory responses to 5-hydroxytryptamine, noradrenaline and acetylcholine injected into the lateral cerebral ventricles of sheep, goats and rabbits. J. Physiol. 212, 377.

Bloom, F.E., Oliver, A.P. & Salmoiraghi, G.C. (1963) Responsiveness of individual hypothalamic neurons to microelectrophoretically administered endogenous amines. Int. J. Neuropharmacol. 2, 181.

Carlsson, A., Falck, B. & Hillarp, N. (1962) Cellular localization of brain monoamines. Acta. Physiol. Scand. 56, Suppl. 196.

Cooper, K.E., Cranston, W.I. & Honour, A.J. (1965) Effects of intraventricular and intrahypothalamic injection of noradrenaline and 5-HT on body temperature in conscious rabbits. J. Physiol. 181, 852.

Cranston, W.I., Hellon, R.F., Luff, R.H. & Rawlins, M.D. (1971) Evidence concerning the effects of endogenous noradrenaline upon body temperature in cats and rabbits. J. Physiol. 212, 24P.

Cross, B.A. & Kitay, J.I. (1967) Unit activity in diencephalic islands. Exp. Neurol. 19, 316.

Cunningham, D.J. Stolwijk, J.A.J., Murakami, N. & Hardy, J.D. (1967) Responses of neurons in the preoptic area to temperature, serotonin and epinephrine. Amer. J. Physiol. 213, 1570.

Dahlström, A. & Fuxe, K. (1964) Evidence for the existence of
    monoamine-containing neurons in the central nervous system. I.
    Demonstration of monoamines in the cell bodies of brain stem
    neurons.   Acta Physiol. Scand. 62, Suppl. 232.

Eccles, J.C., Fatt, P. & Koketsu, K. (1954) Cholinergic and
    inhibitory synapses in a pathway from motor-axon collaterals to
    motoneurones.   J. Physiol. 126, 524.

Euler, C. von (1961) Physiology and pharmacology of temperature
    regulation.   Pharmacol. Rev. 13, 388.

Feldberg, W. (1968) The monoamines of the hypothalamus as mediators
    of temperature responses.   Recent Advances in Pharmacology, ed.
    J.M. Robson & R.S. Stacey. London: Churchill.

Feldberg, W., Hellon, R.F. & Lotti, V.J. (1967) Temperature effects
    produced in dogs and monkeys by injections of monoamines and
    related substances into the third ventricle.   J. Physiol. 191, 501

Feldberg, W., Hellon, R.F. & Myers, R.D. (1966) Effects on
    temperature of monoamines injected into the cerebral ventricles of
    anaesthetized dogs.   J. Physiol. 186, 416.

Feldberg, W. & Myers, R.D. (1964) Effects on temperature of amines
    injected into the cerebral ventricles.   A new concept of
    temperature regulation.   J. Physiol. 173, 226.

Feldberg, W. & Myers, R.D. (1965) Changes in temperature produced by
    microinjections of amines in the anterior hypothalamus of cats.
    J. Physiol. 177, 239.

Feldberg, W. & Myers, R.D. (1966) Appearance of 5-hydroxytryptamine
    and an unidentifed pharmacologically active lipid acid in effluent
    from perfused cerebral ventricles.   J. Physiol 184, 837.

Feldberg, W. & Saxena, P.N. (1971) Fever produced in cats and
    rabbits by prostaglandin E_1 injected into the cerebral
    ventricles.   J. Physiol. 215, 23P.

Findlay, J.D. & Thompson, G.E. (1968) The effect of intraventricular
    injections of noradrenaline, 5-hydroxytryptamine, acetylcholine
    and tranylcypromine on the ox (Bos Taurus) at different
    environmental temperatures.   J. Physiol. 194, 809.

Hoffer, B.J., Siggins, G.R. & Bloom, F.E. (1971) Studies on
    norepinephrine-containing afferents to Purkinje cells of rat
    cerebellum. II.   Sensitivity of Purkinje cells to norepinephrine
    and related substances administered by microiontophoresis.
    Brain Res. 25, 523.

Hulst, S.G.T. & de Wied, D. (1967) Changes in body temperature and
    water intake following intracerebral implantation of carbachol
    in rats.   Physiol. Behav. 2, 367.

Kirkpatrick, W.E., Lomax, P. & Jenden, D.J. (1969), Iontophoretic
    application of acetylcholine to thermoregulatory centers of rats.
    Proc. Western Pharmacol. Soc. 12, 72.

McLennan, H. (1963) Synaptic transmission.   p. 23, Philadelphia:
    Saunders.

Milton, A.S. & Wendlandt, S. (1970) A possible role for
    prostaglandin E_1 as a modulator for temperature regulation in the
    central nervous system of the cat.   J. Physiol. 207, 76P.

Myers, R.D. (1967a) Transfusion of cerebrospinal fluid and tissue bound chemical factors between the brains of conscious monkeys: a new neurobiological assay.   Physiol. Behav. $\underline{2}$, 373.

Myers, R.D. (1967b) Release of chemical factors from the diencephalic region of the unanaesthetized monkey during changes in body temperature.   J. Physiol. $\underline{188}$, 50P.

Myers, R.D. & Beleslin, D.B. (1971) Changes in serotonin release in the hypothalamus during cooling and warming of the monkey. Amer. J. Physiol. $\underline{220}$, 1846.

Myers, R.D. & Sharpe, L.G. (1968) Temperature in the monkey: transmitter factors released from the brain during thermoregulation.   Science $\underline{161}$, 572.

Myers, R.D. & Yaksh, T.L. (1968) Feeding and temperature responses in the unrestrained rat after injections of cholinergic and aminergic substances into the cerebral ventricles.   Physiol. Behav. $\underline{3}$, 917.

Myers, R.D. & Yaksh, T.L. (1969) Control of body temperature in the unanaesthetized monkey by cholinergic and aminergic systems in the hypothalamus.   J. Physiol. $\underline{202}$, 483.

Vogt, M. (1954) The concentration of sympathin in different parts of the central nervous system under normal conditions and after the administration of drugs.   J. Physiol. $\underline{123}$, 451.

EXTRAHYPOTHALAMIC DEEP
BODY THERMOSENSITIVITY

F.W. Klussmann and Fr.-K. Pierau
Kerckhoff-Institut, Max-Planck-Gesellschaft, Bad Nauheim, Germany

About 20 years ago, BAZETT (1951) postulated the existence of what he called, "deep heat receptors". The presumed depth of these "deep receptors" from the surface of the skin was calculated to be 3 - 6 mm, and, for this reason, BAZETT distinguished them from the more superficial lying thermoreceptors. The "deep receptors" were supposed to play a role in the regulation of sweating and were thought to be stimulated by heat produced internally from working muscles. Considering, however, their very small distance from the skin surface, even this type of thermoreceptor might now-a-days be regarded as being still peripheral, belonging functionally to the body shell.

The thermosensitive structures which are dealt with in this article are those of the body core with the exception of those situated in the thermoregulatory centre of the hypothalamus. The latter might be called "central" thermoreceptors to distinguish them from the other deep body thermosensitive structures. Both the hypothalamic and the extrahypothalamic deep body thermoreceptors are thought to measure the temperature of the body core.

Among the first experiments that indicated the possible existence of extracerebral deep body thermosensitivity were those of POPOFF (1934), THAUER (1935), and THAUER and PETERS (1937). In these experiments the nervous connections between the hypothalamus and the spinal cord were severed by a transection of the brain stem or of the spinal cord at the cervical level and the animals were then tested for their ability to maintain body temperature at low ambient temperatures. An example from these experiments is shown in Fig. 1. It is obvious that with increasing time after the spinal transection the animals ability to keep its rectal temperature during exposure to a relatively low ambient temperature has improved considerably.

The explanation for this restitution in thermoregulatory ability, of course, was that thermosensitive structures other than the hypothalamus, but also situated somewhere in the body core had taken over to defend the animal's core temperature against dropping to lethal levels. Support for this hypothesis, that the hypothalamus is not the sole location within the body core which is not only temperature sensitive but also plays a significant role in thermoregulation has been only sporadic for a long time. During the last 15 years, however, evidence has accumulated greatly that an extrahypothalamic deep body thermosensitivity indeed exists.

CHATONNET and TANCHE (1957) concluded from experiments with cooling the body core that the deep body temperature must play an important role in initiating the cold defense mechanisms. BLATTEIS (1960) demonstrated that shivering in the dog when a limb was immersed in cold water did not depend on intact sensory innervation of the limb but on the integrity of its circulation. Since the shivering was independent of changes in the hypothalamic temperature, it was concluded that cold receptors exist in or close to the

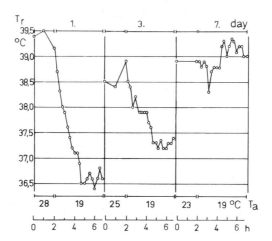

Fig. 1. Time course of rectal temperature of an unanaesthetized rabbit during the exposure to high (28 °C)
and low (19 °C) ambient temperatures (Ta) on the
1st, 3rd, and 7th day after complete spinal transection at the level of C_4. (From THAUER, 1935)

walls of the venous trunk vessels. BLIGH (1963) came to a similar
conclusion from his experiments in which he observed a suppression
of panting in sheep when cold saline was injected into the vena
cava of the animals. Fig. 2 is taken from these experiments. It
shows, how the increase in respiration rate of 10 sheep during exposure to high ambient conditions is influenced by the infusion of
warm or cold saline into the vena cava. Infusion of body-warm saline renders the increase in respiratory rate unchanged. However,
during infusion of cold saline the mean depression of respiration
is obvious although the scatter is very wide. Since there was no
simultaneous decrease of the carotid blood temperature, BLIGH therefore, also proposed the existence of cold sensitive elements in or
near the walls of the vena cava and other structures of central
circulation. From his experiments on temperature regulation in exercise ROBINSON (1963) suggested, that thermoreceptors, capable of
reflexly exciting the sweat glands, might be located in the veins
which drain warm blood from the working muscles. DOWNEY and coworkers (1964), too, considered the existence of thermosensitive structures in the heart or great vessels.

    Evidence for another site of deep body thermosensitivity has
been given recently by RAWSON and QUICK (1971). They made it probable that temperature sensitive receptors are located in the abdominal viscera of the ewe. Selective heating of this region could significantly modify a steady state thermoregulatory drive.

    In many of the experiments with cooling or warming extrahypotha-
lamic parts of the body core as for example those of CHATONNET and
TANCHE (1957), the possibility still existed that besides the deep
body thermosensitive structures thermoreceptors of the skin also
might have been stimulated. The influence of the skin temperature
receptors, however, was excluded in the experiments of HALLWACHS
and coworkers (1961) and RAUTENBERG and coworkers (1963). In these
experiments the supraspinal structures of dogs were kept at con-

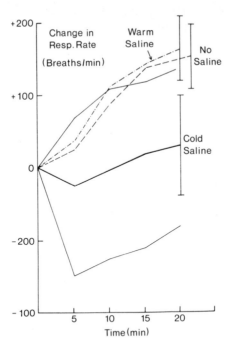

Fig. 2. Change in respiratory rate of 10 sheep during ex-
        posure to high ambient temperature and humidity.
        (Modified after BLIGH, 1963) - - - - Uninfluenced
        animals (mean values with s.d.) -·-·-· During
        simultaneous infusion of body-warm saline (39 - 40 °C)
        (mean values with s.d.) ——— During simultaneous
        infusion of cold saline (20 - 28 °C). (Thin lines:
        extreme values; thick line: mean values with s.d.)

stant temperatures by heat exchangers in both carotid arteries and
both jugular veins. The cold receptors of the skin were prevented
from being stimulated by immersing the animal completely in a warm
water bath of 36 °C except for the mouth and nasal region. Lowering
the temperature of the deep body core by means of a thermode placed
in the esophagus and stomach and by cooling the blood flowing back
in the jugular veins led to strong shivering and to a considerable
increase in oxygen consumption. Very obviously, this increase in
heat production which could only be due to lowering deep body temperature, conclusively pointed to the existence of deep body thermosensitive structures.

    Of course, the immediate question arose as to where these struc-
tures might be located. It could well be that the thermoreceptive
structures in or near the walls of the great veins as postulated by
BLATTEIS (1960) and BLIGH (1963) are responsible for these reactions. Another possibility are the muscle spindles of peripheral
muscles. The thermosensitivity of these structures and the close
resemblance of their discharge patterns with those of cold receptors had been shown by LIPPOLD and coworkers (1960). On the other
hand, it is not likely that the thermosensitive structures within
the abdominal viscera described by RAWSON and QUICK (1971) are the

mediators, since these receptors are thought to respond only to
heating and not to cooling.

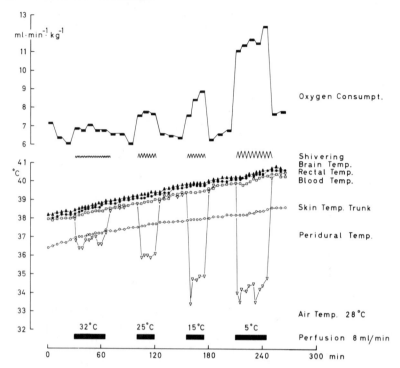

Fig. 3. Effects of isolated cooling of the spinal cord of
an anaesthetized dog on shivering, oxygen consumption, and various body temperatures. The black bars
indicate the cooling periods with increasing intensity. (From SIMON et al., 1964)

Another avenue of approach was opened with the experiments by
SIMON et al. (1964), in which they were able to elicit typical cold
shivering and vasoconstriction by selectively cooling within the
vertebral canal. Fig. 3 shows one of their experiments. It is obvious that with increasing "cold load" the temperature within the
vertebral canal is lowered more and more. With the decrease in
spinal temperature shivering is generated and the heat production
rises, as measured by the oxygen consumption. These effects were
not suppressed by the rising temperature of the brain, the blood
and the rectum and the cutaneous temperature. In another series of
experiments, already existing shivering due to peripheral cooling
could be enhanced by additional cooling of the spinal cord. Selective warming of the spinal cord, on the other hand, suppressed already existing shivering and induced a cutaneous vasodilatation
(RAUTENBERG and SIMON, 1964). From these experiments the conclusion
was drawn, that within the spinal cord or within the vicinity of
the vertebral canal cold and warm sensitive structures must exist
which are responsible for mediating the thermoregulatory reactions
like shivering, vasoconstriction and vasodilatation.

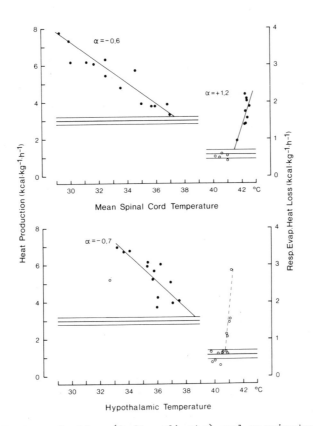

Fig. 4. Heat production (left ordinate) and respiratory
evaporative heat loss (right ordinate) related
to spinal cord temperature (above) and hypothalamic temperature (below). Mean values and standard deviations are given by the three parallel
lines. Only filled circles were used for calculation of regression lines. 56 stimulation
periods in one dog at an ambient air temperature
of 18 °C. (From JESSEN and MAYER, 1971)

These results, of course, provoked the question if the other
heat dissipating mechanisms like panting and sweating can also be
initiated by warming the spinal cord selectively. That this is indeed the case, was shown by JESSEN and MAYER (1971). In these experiments the responses to isolated cooling or warming of the spinal cord and hypothalamus in trained unanaesthetized dogs were
compared on a quantitative basis. Fig. 4 shows the results of 56
such thermal stimulations in 7 experiments performed on one dog at
an ambient air temperature of 18 °C. The upper part of the figure
shows the results to spinal cord cooling and warming, the lower
those of hypothalamic cooling and warming. Each circle represents
one stimulation. It is obvious from this figure that cooling either
the spinal cord or the hypothalamus produced about the same increase in heat production, the proportionality constants for both

regression lines being almost identical. The proportionality constant for respiratory heat loss during spinal cord heating was a-
bout twice that of cooling. Probably due to the low ambient temperature, only five periods of hypothalamic heating led to an augment-
ed respiratory heat loss. At higher ambient temperatures, however,
hypothalamic heating was more successful.

From these experiments and from those with simultaneous concomitant and opposite temperature changes in the spinal cord and
hypothalamus JESSEN and coworkers (1971) concluded that the temperature signals from both areas are added to give a combined drive
to the effector systems, and that both the hypothalamus and the
spinal cord represent basically equivalent sensors of body core
temperature. In another set of experiments with oxen McLEAN et al.
(1970) revealed that isolated warming of the spinal cord leads to
an increase in the cutaneous moisture loss of the animal.

KOSAKA et al. (1969) demonstrated that thermal stimulation of the
spinal cord has an effect on respiratory and cortical activity of
the rabbit, and that the respiratory responses to spinal cord warming or cooling are mediated by diencephalic structures. BRÜCK and
WÜNNENBERG (1967) in their experiments on the regulation of shivering in guinea pigs were able to suppress by spinal cord warming
shivering which had been induced by hypothalamic cooling. From
these results it was concluded that ascending signals from the
thermosensitive structures in the spinal cord must be integrated in
the hypothalamus. Indeed, in their recordings from the spinothalamic tract WÜNNENBERG and BRÜCK (1970) found ascending units which
increased in firing rate when the spinal cord was selectively warm-
ed in the region between C_5 and T_1. In addition to these warm-
sensitive units SIMON and IRIKI (1971) have described units in the
spinothalamic tract of cats responding to cooling the spinal cord.
In their experiments the number of warm-sensitive units was about
five times greater than those of the cold units.

That integration at the hypothalamic level of temperature signals arising in the spinal cord and the hypothalamus does in fact
take place has been shown by GUIEU and HARDY (1970). Fig. 5 is
taken from their experiments. It demonstrates that selective warming of the preoptic area of the hypothalamus as well as isolated
warming of the spinal cord increases the firing rate of a preoptic
neurone and the respiratory frequency. At the same time the ear
temperature increased with heating either area. A similar convergence of temperature signals from the spinal cord and the preoptic
temperature sensitive region upon neurones in the posterior hypothalamus has been demonstrated by WÜNNENBERG and HARDY (1971).

As indicated earlier and as is obvious from Fig. 5 the temperature changes within the vertebral canal also induce changes in local blood flow of the skin, in this manner decreasing or increasing heat dissipation. That these changes in skin blood flow are
accompanied by appropriate reverse changes in the intestinal area
has been shown in anaesthetized dogs by KULLMANN et al. (1970).
These antagonistic changes of blood flow with spinal cord cooling
and warming are caused by concomitant but antagonistic changes in
cutaneous and visceral sympathetic activity as has been found by
WALTHER et al. (1970).

Since the first description of the thermoregulatory potency of
the spinal cord by SIMON and coworkers (1964) a great number of
papers have appeared which all support this concept. All warm-

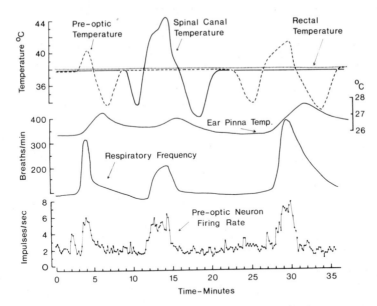

Fig. 5. Influence of preoptic temperature and spinal canal
temperature on ear pinna temperature, respiratory
frequency and firing rate of a preoptic neurone
of a rabbit. (Modified after GUIEU and HARDY, 1970)

blooded experimental animals which have been investigated so far
have shown this spinal thermosensitivity. These include dogs
(JESSEN et al., 1971), cats (KLUSSMANN, 1969), rabbits (KOSAKA and
SIMON, 1968 a,b), guinea pigs (BRÜCK and WÜNNENBERG, 1967), goats
(JESSEN, 1971), pigs (INGRAM and LEGGE, 1971), and oxen (McLEAN et
al., 1970). Even the pigeon with a separate evolutionary development from mammals also has this spinal thermosensitivity, as first
shown by RAUTENBERG (1969).

All these observations on various species naturally led to the
assumption that the phenomena produced by local temperature changes
within the vertebral canal might be caused by a temperature change
of the spinal cord itself. This assumption was supported by the
finding of MEURER and coworkers (1967), that in chronically deaf-
ferented dogs shivering can be induced by spinal cord cooling. In
addition, a great number of experiments with acutely and chronically spinalized dogs and rabbits by SIMON et al. (1966), KOSAKA and
SIMON (1968 a,b) and SIMON (1969) made it clear, that shivering
and vasoconstriction of the skin can also be elicited in a spinalized animal by selectively cooling the spinal cord. These experiments, therefore, left no doubt, that the reactions observed during
cooling of the vertebral canal, must originate in the spinal cord
itself. Furthermore, they proved that the spinal cord is equipped
with all the mechanisms necessary for the production of shivering.

The question then arises: what are the thermosensitive structures within the spinal cord?
From the classical thermoregulatory point of view one might expect
that the spinal cord contains specific cold and warm receptors.

However, sofar no such specific thermoreceptive structures within
the spinal cord have been described. The thermosensitivity of ascending fibres in the spinothalamic tract, as described by WÜNNEN-
BERG and BRÜCK (1970) and SIMON and IRIKI (1971) could well be
caused by a thermal sensitivity of the relaying second order neurons for these fibres in the dorsal columns.

Another point of view would be that many if not all neuronal
structures of the spinal cord possess such a thermosensitivity,
that e.g. cooling leads to an increased ascending activity, increased vasomotor tone of the skin, and increased motor output and
by this to an increased heat production. Evidence that many parts
of the central nervous system are thermosensitive was demonstrated
for the sensorimotor cortex by GARTSIDE and LIPPOLD (1967) and more
recently by BARKER and CARPENTER (1970). On the basis of their
findings, BARKER and CARPENTER (1971) also cautioned "to imply that
thermosensitivity (of neurones) indicate a role in the transmission
of thermal information or in thermoregulation". This objection, of
course, holds also true for the thermosensitivity of hypothalamic
neurones. That other motor systems, not particularly involved in
thermoregulation can also be activated by cooling them locally was
shown by PIERAU et al. (1970) for the striated pupil muscle of
pigeons. Local cooling of the oculomotor nucleus led to an increased activity of the iris sphincter muscle. Whole-body-cooling of
dogs was followed by an increase in integrated phrenic nerve activity (PLESCHKA, 1969), which is mainly caused by a prolonged firing
of the individual phrenic motoneurone and by the recruitment of new
units through the fall in temperature (BOCK and PLESCHKA, 1971).

In order to test the assumption, that the efferent systems of
the spinal cord themselves might be thermosensitive in the sense
that this thermosensitivity contributes to the observed thermoregulatory reactions during isolated spinal cord cooling and warming,
single fibre recordings and intracellular recordings from the efferent neurones of the spinal cord were required. The system which
is easiest to approach for this purpose and which is also heavily
involved in thermoregulatory processes is the motor system of the
spinal cord. Single fibre recordings from ventral and dorsal roots
of cats revealed that during isolated cooling of the spinal cord
the increase in γ-motoneurone activity always preceded the activation of α-motoneurones and the appearance of shivering (KLUSSMANN,
1969). Within the group of α-motoneurones the smaller tonic α-cells
were usually activated after a smaller drop in spinal temperature
than the bigger α-motoneurones of the phasic type (STELTER and
KLUSSMANN, 1969). These findings agreed with the observation that
spinal cord cooling rapidly leads to a decrease in reflex-tension
of the "red" M. soleus, mainly innervated by tonic fibres. In the
"pale" muscles, however, (M. tibialis ant. and M. extensor digit.
long.), which have a high proportion of phasic innervation, the
reflex-tension increased to a maximum at extraspinal temperatures
between 35 °C and 30 °C (STELTER et al., 1969). Recordings of the
stretch responses of primary and secondary muscle spindle endings
of various hind-leg muscles also revealed that cooling of the
spinal cord leads to a decrease in dynamic sensitivity of mainly
the primary muscle spindle endings (KLUSSMANN and HENATSCH, 1969).

These experiments suggested that the differential thermosensitivity of the motor cells somehow depended on cell size or some factor correlated with size. It was also tentatively concluded that
the spinal motoneurones themselves might be thermosensitive. However, the increased output of motor cells with cooling the spinal

cord could still be caused by some other neuronal structures presynaptic to the motoneurones under investigation. Increased driving
of motor cells at lowered spinal temperatures via presynaptic interneurones had been suggested by BROOKS et al. (1955). To prove
the hypothesis that the motoneurones themselves are thermosensitive
in such a way that a fall in their temperature leads to an increased output, intracellular recordings from spinal motoneurones were
done by PIERAU et al. (1969). It was also hoped that these cells
might serve as models for the thermoreceptive structures in other
parts of the CNS, e.g. the hypothalamus, since it seemed reasonable
to assume that the temperature dependent changes in membrane characteristics in hypothalamic and other thermosensitive neurones
might be basically the same as those in the spinal cells.

Fig. 6 gives two examples from the intracellular recordings made
by PIERAU (1971). In part A the motoneurone of the lumbar region
was stimulated antidromically over the ventral root. On a first

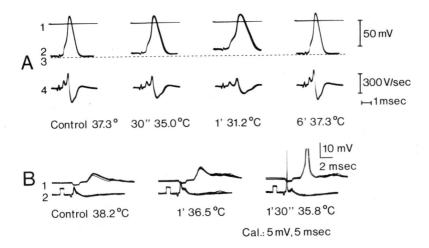

Fig. 6. Effects of local spinal cord cooling on intra-
        cellularly recorded membrane potentials, action
        potentials, and excitatory postsynaptic potentials
        (ESP) of two lumbar motoneurones of the cat. In
        each recording 8 - 10 traces superimposed. Time
        after onset of cooling; intraspinal temperatures.
        (Modified after PIERAU et al., 1969)
        A: Antidromic stimulation of ventral root L_7
        1: Reference line for zero membrane potential;
        2: Intracellular recording;
        3: Reference line for control value of resting
           membrane potential;
        4: First derivative of recorded spike.
        B: Orthodromic stimulation of dorsal root S_1
        1 and 2: Intracellularly recorded membrane and
           action potentials with two different ampli-
           fications and sweep speeds (spikes are
           truncated).

glance, only the negative effects, that is the depressive effects of lowering the spinal temperature on form and time course of the action potentials are obvious. The rate of depolarization and re-polarization slows down and the spike duration increases. This is especially evident from the first derivatives. These changes are reversible with warming. At closer inspection, however, a small but significant depolarization of the membrane potential with the fall in intraspinal temperature can be seen, most typically between the last two recordings of part A of Fig. 6. Such a depolarization means that the cell is driven closer to its threshold. Hence, the fall in intraspinal temperature produces not only depressant changes in spike properties as one would expect from the Arrhenius equation, but also a characteristic depolarization which tends to increase the excitability of the cell.

Another change in membrane characteristics that also results in an increased excitability with decreasing temperature and therefore supplements the simultaneous depolarization is shown in part B of Fig. 6. An orthodromic stimulus over the dorsal root was adjusted in its strength such that a typical excitatory postsynaptic poten-tial, an EPSP, is generated in this cell, which however, at normal temperature, remained below threshold. With cooling the amplitude of the EPSP increases considerably and at an intraspinal tempera-ture of 35,8 °C every EPSP reaches threshold and triggers a spike although the stimulus strength has not been changed.

Fig. 7 demonstrates the high thermosensitivity of these changes and that they are all reversible with rewarming. In this case two stimuli were adjusted in their strength such that two EPSPs partly summed, but at normal temperature, remained below threshold.

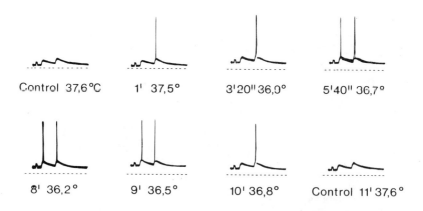

Control 37,6°C   1' 37,5°   3'20" 36,9°   5'40" 36,7°

8' 36,2°   9' 36,5°   10' 36,8°   Control 11' 37,6°

Cal.: 5 mV, 2 msec

Fig. 7. Effects of local spinal cord cooling on intracellularly recorded membrane potential, action
potentials, and excitatory postsynaptic potentials
(EPSP) of a lumbar motoneurone of the cat. Double
shock orthodromic stimulation of biceps-semitendi-
nosus nerve. In each recording 8 - 10 traces
superimposed. Time after onset of cooling; intraspinal temperatures. (From PIERAU, 1971)

After one minute of cooling the amplitudes of both EPSPs have increased so much that a few of the 8 - 10 second EPSPs reach the
threshold and a spike is triggered. With further cooling each of
the second EPSPs fires the cell and then even the first EPSPs reach
threshold and generate a spike. With rewarming these changes are
reversible. The high thermosensitivity of these effects is very remarkable. Even a change of only a tenth of a degree can produce
measurable increases in excitability of the cell, which then become
very obvious after a drop in intraspinal temperature of about a degree centigrade. All these changes in excitability occur within a
temperature range in which normal temperature regulation processes
happen.

The same holds true for the effects of warming the spinal cord
above the normal temperature. Fig. 8 gives the reactions of two
motoneurones to warming the spinal cord. In part A of Fig. 8 the
strength of the afferent volley was adjusted such that at normal
temperature every EPSP generated an action potential. It is obvious
that one and a half minutes after onset of warming at an intraspinal
temperature of 38.8 °C some of the EPSPs already fail to reach
threshold. After a further increase in temperature by two tenths of
a degree almost all of the EPSPs remain subthreshold. This decrease
in excitability is probably due to a hyperpolarization which cannot
be seen from Fig. 8 because the oscilloscope beam was held in a
fixed position, but which has been found by PIERAU (1971) in many
other recordings. The other reason for the failing of the cell to
reach threshold is the diminished size of the EPSP.

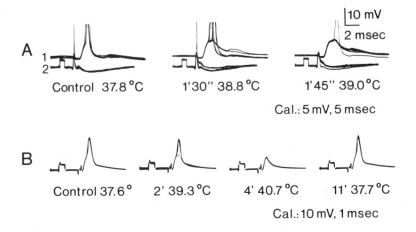

Fig. 8. Effects of local spinal cord warming on intra-
        cellularly recorded membrane potential, action
        potentials, and excitatory postsynaptic potentials
        (EPSP) of two lumbar motoneurones of the cat. In
        each recording 8 - 10 traces superimposed. Time
        after onset of warming; intraspinal temperatures.
        (Modified after PIERAU et al., 1969)
        A: Orthodromic stimulation of dorsal root S_1
        1 and 2: Intracellularly recorded potentials
            with two different amplifications and sweep
            speeds. Spikes are truncated.
        B: Antidromic stimulation of ventral root L_7.

The recordings in part B of Fig. 8 show the effects of warming in another motoneurone, this time using an antidromic stimulation. Under these conditions, of course, no EPSP is generated, but the cell is fired from the impulse over its own axon. Warming the spinal cord apparently makes the cell more resistive to the invading impulse since some of the stimulations fail to trigger a full-blown spike. At an intraspinal temperature of 40.7 °C only the so-called initial segment portion of the spike remains, while the so-called somadendritic part is missing. This is due to the higher threshold of the somadendritic portion of cells in comparison to the lower threshold of their initial segment where the cell-axon originates (ECCLES, 1957). Any suppression in excitability through the increased spinal temperature must, therefore, first appear in the somadendritic part of the action potential.

The reason for the decrease or increase in membrane potential with decreasing or increasing temperature could still very well be due to modified interneuronal activity presynaptic to the motoneurones (BROOKS et al., 1955). The same could be true for changes in size of polysynaptic EPSP with temperature. Another factor, however, which influences greatly the EPSP-size and therefore the susceptibility of the cell to discharge is the resistance of the cell membrane. The fact, that the size of monosynaptic EPSP also increases with cooling had indicated a change of membrane resistance with temperature. Fig. 9 demonstrates as an example, how cooling the spinal cord affects the membrane resistance of a lumbar α-motoneurone. In the upper row the effects are shown of lowering the intraspinal temperature on stimulation the cell by injecting a current pulse through the microelectrode. At normal temperature the size of the current injection was adjusted such that the resulting depolarization was still subthreshold. By lowering the spinal cord temperature, with the same current injections the depolarizations became supra-threshold and a spike was generated. Under these experimental conditions of current injection, the change in membrane potential during cooling can only be caused by a change in membrane resistance. In order to measure variations in membrane resistance with changing temperature, current of positive or negative polarity was injected and the resultant alterations in membrane voltage measured. The insets in Fig. 9 show the current injections and the resulting potential changes. The plot of calculated membrane resistance clearly demonstrates an increase in membrane resistance with decreasing temperature, and the return to precooling level with the return to normal temperature. Warming the spinal cord above normal temperature (not shown in Fig. 9) had the opposite effect, the membrane resistance decreased with increasing temperature (PIERAU, 1971). This change in membrane resistance with cooling or warming is an alteration in one of the membrane properties at the postsynaptic membrane and cannot be caused by presynaptic events. The observed variations in the size of the EPSP are also probably caused mainly by this change in membrane resistance. A contribution to the change in EPSP amplitude by presynaptic influences, e.g. activation of excitatory interneurones, seems reasonable. However, the relative proportion of such a presynaptic contribution is difficult to estimate.

If the change in EPSP amplitude with cooling or warming is caused by changes in membrane resistance, then the size of the opposite postsynaptic potential changes, that is the size of inhibitory postsynaptic potential (IPSP) should also vary with temperature.

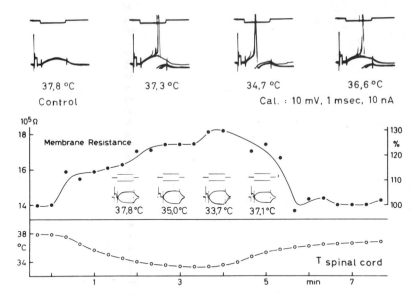

Fig. 9. Effects of local spinal cord cooling on intracellularly evoked and recorded action potentials
of a cat lumbar motoneurone (uppermost recording:
intracellularly applied depolarizing current) and
time course of calculated membrane resistance and
intraspinal temperature. Inset: Depolarizing and
hyperpolarizing current pulses (upper recordings)
and resultant intracellular potential changes
(lower recordings) at various intraspinal
temperatures. (Modified from PIERAU et al., 1969)

In part A of Fig. 10 are shown the effects of cooling on the IPSP
in a lumbar motoneurone after stimulating the dorsal root S_1. With
lowering the temperature the size of the IPSPs increases, as one
would have expected from the concomitant increase in membrane resistance. The opposite effect happens with increasing spinal temperature, the size of the IPSP becomes less as shown in part B of
Fig. 10.

Inspite of the increase in IPSP-size with spinal cooling, in
many recordings the simultaneous increase of the EPSPs superseded
the increase in IPSP amplitude and the cell still discharged. The
question then, if a motoneurone discharges or not during cooling or
warming, depends on the balance between excitatory and inhibitory
influences. Fig. 11 shows in a very schematic manner the various
thermoregulatory influences on a motoneurone. Excluding all other
non-thermal exciting or inhibiting influences from muscles, skin,
joints, segmental and supraspinal sources, and focussing the attention on the thermoregulatory important synaptic connections only,
it can be said that the motoneurone is of course influenced by des-
cending fibre tracts from the thermoregulatory centres of the
hypothalamus and perhaps from the reticular formation of the brain
stem, which has been found thermosensitive (NAKAYAMA and HARDY,
1969). These connections may be excitatory for initiating cold defense reactions or they may be inhibiting, minimizing heat produc-

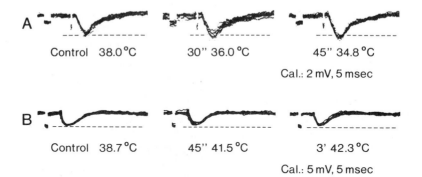

A Control 38.0 °C   30" 36.0 °C   45" 34.8 °C

Cal.: 2 mV, 5 msec

B Control 38.7 °C   45" 41.5 °C   3' 42.3 °C

Cal.: 5 mV, 5 msec

Fig. 10. Effects of local spinal cord cooling (A) and
warming (B) on intracellularly recorded membrane
potentials and inhibitory postsynaptic potentials
(IPSP) of two lumbar motoneurones of the cat. In
each recording 8 - 10 traces superimposed. Orthodromic stimulation of dorsal root S_1. Time after
onset of cooling or warming; intraspinal temperatures. (From PIERAU, 1971)

tion depending on the actual temperature of the hypothalamus, its
set point (HAMMEL, 1970), or its gain (MITCHELL, 1970). The motoneurone is also driven with certainty by the specific intraspinal
thermosensitive structures, if they exist. Again we might need an
excitatory and inhibitory input. And finally the neurone is probably somehow connected at the spinal level with the specific thermoreceptors of the other areas of the body core and the skin, here
too one probably has to distinguish between warm and cold receptors. In addition, considering the evidence, presented in this
review, we have to conclude that the motoneurone itself is thermosensitive in such a way that a drop in its temperature leaves it
more excitable and, vice versa, a rise in its temperature renders
it less excitable. This means that the gain of the neurone for all
incoming signals changes with its temperature. In addition, the
thermosensitivity of the motoneurones includes the necessary negative feedback, since the fall in its temperature increases its output, and by this means increases heat production. Any theory about
thermoregulatory networks will have to take into account this
thermosensitivity. This seems inevitable since of all the some 130
or so motoneurones which have been tested so far, about 90 % reveal
the described thermosensitivity (PIERAU, 1971). For one major aspect, therefore, of thermoregulation, namely the heat production
through shivering and to this aspect only, the motoneurone of the
spinal cord suggests itself as the final integrating system for the
many thermoregulatory signals from a variety of thermosensitive
areas of the body core and body shell and for its own temperature.

The question remains how much in a quantitative way this direct thermosensitivity of spinal motoneurones contributes to the increased motor output during isolated spinal cord or whole body cooling.

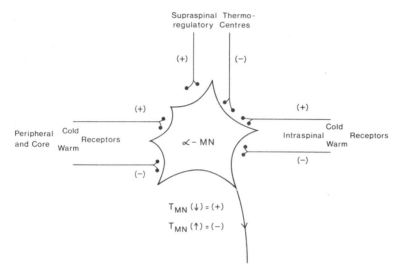

Fig. 11. Schematic diagram for the integrative function of spinal motoneurones for the activating (+) and inactivating (-) inputs from various thermosensitive structures and for its own temperature. T_{MN} ($\uparrow$) and T_{MN} ($\downarrow$) = increase or decrease of the local temperature of the motoneurone.

Although, as yet, no appropriate changes in membrane characteristics of motoneurones have been found to explain the behaviour of warm-sensitive neurones in the spinal cord or hypothalamus it is possible that a different synaptic organization, different temperature coefficients for membrane potential and membrane resistance and different permeabilities for K^+ or Na^+ are responsible for the warm-sensitivity of these neurones. Extracellular recordings from Renshaw-cells, ascending neurones, and other interneurones revealed an increase in firing rate with heating the spinal cord and a decrease with cooling. Intracellular recordings of membrane properties in these cells will be necessary to establish the differences between cold- and warm-sensitive neurones.

S U M M A R Y

The evidence for extrahypothalamic deep body thermosensitivity has accumulated greatly in recent years. Locations for these thermoreceptors might be the walls of blood vessels, especially the great veins, the muscle spindles of the muscles or the walls of various intestinal organs. Many experimental results also point to the spinal cord. Its proper thermosensitivity has been shown for a great number of mammals and even for birds. Selective cooling or warming the spinal cord produces every thermoregulatory defense reaction, known from the stimulation of other thermoreceptive areas such as the skin or the hypothalamus. Within the spinal cord ascending neurones exist which are sensitive to both cold and warm. Cold reactions like shivering and vasoconstriction can, on principle, be generated by the spinal cord without supraspinal control.

The membranes of α-motoneurones are the only neurones in warm-blooded animals which so far have been investigated by intracellular methods in vivo for the thermosensitivity of their membrane properties. They have been found to possess the proper thermosensitivity partly to explain the motor reactions to selective cooling or warming of the spinal cord. Furthermore, the motoneurones of the spinal cord which with cooling had shown an increase in excitability and discharge frequency may serve as a model for other cold-sensitive neurones in the spinal cord, the hypothalamus and the other thermosensitive structures of the body core, since it is quite conceivable that the same temperature dependent changes in membrane properties will be found in these neurones.

R E F E R E N C E S

BARKER, J.L., and D.O. CARPENTER (1970). Thermosensitivity of neurons in the sensimotor cortex of the cat. Science 169, 597.
BARKER, J.L., and D.O. CARPENTER (1971). Neuronal thermosensitivity. Science 172, 1361.
BAZETT, H.C. (1951). Theory of reflex controls to explain regulation of body temperature at rest and during exercise. J.appl. Physiol. 4, 245.
BLATTEIS, C.M. (1960). Afferent initiation of shivering. Amer.J. Physiol. 199, 697.
BLIGH, J. (1963). The receptors concerned in the respiratory response to humidity in sheep at high ambient temperature. J.Physiol.(Lond.) 168, 747.
BOCK, K.H., und K. PLESCHKA (1970). Das Verhalten der Motoneurone des N. phrenicus in Hypothermie. Pflügers Arch.ges.Physiol. 316, R 55.
BROOKS, C.McC., K. KOIZUMI and J.L. MALCOLM (1955). Effects of changes in temperature on reactions of spinal cord. J.Neurophysiol. 18, 205.
BRÜCK, K., und W. WÜNNENBERG (1967). Die Steuerung des Kältezitterns beim Meerschweinchen. Pflügers Arch.ges.Physiol. 293, 215.
CHATONNET, J., et M. TANCHE (1957). Arguments en faveur d'une regulation centrale de la thermogénèse chez le chien. J.Physiol. (Paris) 49, 89.
DOWNEY, J.A., R.F. MOTTRAM and G.W. PICKERING (1964). The location by regional cooling of central temperature receptors in the conscious rabbit. J.Physiol.(Lond.) 170, 415.
ECCLES, J.C. (1957). The physiology of nerve cells. Baltimore: The Johns Hopkins Press.
GARTSIDE, J.B., and O.C.J. LIPPOLD (1967). Production of persistent changes in level of neuronal activity by brief local cooling of cerebral cortex of rat. J.Physiol.(Lond.) 189, 475.
GUIEU, J.D., and J.D. HARDY (1970). Effects of heating and cooling of the spinal cord on preoptic unit activity. J.appl.Physiol. 29, 675.
HALLWACHS, O., R. THAUER und W. USINGER (1961). Die Bedeutung der tiefen Körpertemperatur für die Auslösung der chemischen Temperaturregulation. II. Kältezittern durch Senkung der tiefen Körpertemperatur bei konstanter, erhöhter Haut- und Hirntemperatur. Pflügers Arch.ges.Physiol. 274, 115.

HAMMEL, H.T. (1970). Concept of the adjustable set temperature.
    In: Hardy, J.D., A.Ph. Gagge and J.A.J. Stolwijk (Editors)
    "Physiological and Behavioral Temperature Regulation", Spring-
    field, Ill., Charles C. Thomas.
INGRAM, D.L., and K.F. LEGGE (1971). The influence of deep body
    temperatures and skin temperatures on peripheral blood flow in
    the pig. J.Physiol.(Lond.) 215, 693.
JESSEN, C. (1971). Personal communication.
JESSEN, C., and E.Th. MAYER (1971). Spinal cord and hypothalamus as
    core sensors of temperature in the conscious dog. I. Equivalence
    of responses. Pflügers Arch.ges.Physiol. 324, 189.
JESSEN, C., and O. LUDWIG (1971). Spinal cord and hypothalamus as
    core sensors of temperature in the conscious dog. II. Addition
    of signals. Pflügers Arch.ges.Physiol. 324, 205.
JESSEN, C., and E. SIMON (1971). Spinal cord and hypothalamus as
    core sensors of temperature in the conscious dog. III. Identity
    of functions. Pflügers Arch.ges.Physiol. 324, 217.
KLUSSMANN, F.W. (1969). Der Einfluß der Temperatur auf die afferen-
    te und efferente motorische Innervation des Rückenmarks. I.
    Temperaturabhängigkeit der afferenten und efferenten Spontan-
    tätigkeit. Pflügers Arch.ges.Physiol. 305, 295.
KLUSSMANN, F.W., und H.D. HENATSCH (1969). Der Einfluß der Tempera-
    tur auf die afferente und efferente motorische Innervation des
    Rückenmarks. II. Temperaturabhängigkeit der Muskelspindelfunk-
    tion. Pflügers Arch.ges.Physiol. 305, 316.
KOSAKA, M., und E. SIMON (1968 a). Kältetremor wacher, chronisch
    spinalisierter Kaninchen im Vergleich zum Kältezittern intakter
    Tiere. Pflügers Arch.ges.Physiol. 302, 333.
KOSAKA, M., und E. SIMON (1968 b). Der zentralnervöse spinale Me-
    chanismus des Kältezitterns. Pflügers Arch.ges.Physiol. 302, 357.
KOSAKA, M., E. SIMON, R. THAUER and O.-E. WALTHER (1969). Effect of
    thermal stimulation of spinal cord on respiratory and cortical
    activity. Amer.J.Physiol. 217, 858.
KULLMANN, R., W. SCHÖNUNG and E. SIMON (1970). Antagonistic changes
    of blood flow and sympathetic activity in different vascular
    beds following central thermal stimulation. I. Blood flow in
    skin, muscle and intestine during spinal cord heating and cool-
    ing in anaesthetized dogs. Pflügers Arch.ges.Physiol. 319, 146.
LIPPOLD, O.C.J., J.G. NICKOLS and J.W.T. REDFARN (1960). A study
    of the afferent discharge produced by cooling a mammalian muscle
    spindle. J.Physiol.(Lond.) 153, 218.
McLEAN, J.A., J.R.S. HALES, C. JESSEN and D.T. CALVERT (1970). In-
    fluences of spinal cord temperature on heat exchange of the ox.
    Proc.Aust.Physiol.Pharmacol.Soc. 1, 32.
MEURER, K.-A., C. JESSEN und M. IRIKI (1967). Kältezittern während
    isolierter Kühlung des Rückenmarks nach Durchschneidung der
    Hinterwurzeln. Pflügers Arch.ges.Physiol. 293, 236.
MITCHELL, D., J.W. SNELLEN and A.R. ATKINS (1970). Thermoregulation
    during fever: change of set-point or change of gain. Pflügers
    Arch.ges.Physiol. 321, 293.
NAKAYAMA, T., and J.D. HARDY (1969). Unit responses in the rabbit's
    brain stem to changes in brain and cutaneous temperature. J.appl.
    Physiol. 27, 848.
PIERAU, Fr.-K., M.R. KLEE and F.W. KLUSSMANN (1969). Effects of
    local hypo- and hyperthermia on mammalian spinal motoneurones.
    Fed.Proc. 28, 1006.
PIERAU, Fr.-K., E. ALEXANDRIDIS, G. SPAAN, A. OKSCHE und F.W.
    KLUSSMANN (1970). Der Einfluß von lokalen Temperaturänderungen
    im pupillomotorischen Kerngebiet der Taube auf die Aktivität der
    Irismuskulatur. Pflügers Arch.ges.Physiol. 315, 291.

PIERAU, Fr.-K. (1971). Habilitationsschrift. Gießen.
PLESCHKA, K. (1969). Der Einfluß der Temperatur auf die elektrische
    Aktivität des Nervus phrenicus. Pflügers Arch.ges.Physiol. 308,
    333.
POPOFF, N.F. (1934). Die vegetativen Funktionen des Hundes nach
    weitgehender Ausschaltung der Einflüsse des Zentralnervensystems.
    Pflügers Arch.ges.Physiol. 234, 137.
RAUTENBERG, W., E. SIMON und R. THAUER (1963). Die Bedeutung der
    Kerntemperatur für die chemische Temperaturregulation beim Hund
    in leichter Narkose. I. Isolierte Senkung der Rumpfkerntempera-
    tur. Pflügers Arch.ges.Physiol. 278, 337.
RAUTENBERG, W., und E. SIMON (1964). Die Beeinflussung des Kälte-
    zitterns durch lokale Temperaturänderung im Wirbelkanal.
    Pflügers Arch.ges.Physiol. 281, 332.
RAUTENBERG, W. (1969). Die Bedeutung der zentralnervösen Thermo-
    sensitivität für die Temperaturregulation der Taube. Z.vergl.
    Physiol. 62, 235.
RAWSON, R.O., and K.P. QUICK (1971). Thermoregulatory responses to
    temperature signals from the abdominal viscera of sheep. J.
    Physiol. (Paris) 63, 399.
ROBINSON, S. (1963). Temperature regulation in exercise. Pediatrics
    (Suppl.) 32, 691.
SIMON, E., W. RAUTENBERG, R. THAUER und M. IRIKI (1964). Die Aus-
    lösung von Kältezittern durch lokale Kühlung im Wirbelkanal.
    Pflügers Arch.ges.Physiol. 281, 309.
SIMON, E., F.W. KLUSSMANN, W. RAUTENBERG und M. KOSAKA (1966).
    Kältezittern bei narkotisierten spinalen Hunden. Pflügers Arch.
    ges.Physiol. 291, 187.
SIMON, E. (1969). Kreislaufwirkungen der spinalen Hypothermie.
    J.Neuro-Viscer.Relat. 31, 223.
SIMON, E., and M. IRIKI (1971). Sensory transmission of spinal heat
    and cold sensitivity in ascending spinal neurons. Pflügers Arch.
    ges.Physiol. 328, 103.
STELTER, W.-J., und F.W. KLUSSMANN (1969). Der Einfluß der Rücken-
    markstemperatur auf die Dehnungsantwort tonischer und phasischer
    α-Motoneurone. Pflügers Arch.ges.Physiol. 309, 310.
STELTER, W.-J., G. SPAAN und F.W. KLUSSMANN (1969). Der Einfluß
    der spinalen und peripheren Temperatur auf die Reflexspannung
    "roter" und "blasser" Muskeln. Pflügers Arch.ges.Physiol. 312, 1.
THAUER, R. (1935). Wärmeregulation und Fieberfähigkeit nach opera-
    tiven Eingriffen am Nervensystem homoiothermer Säugetiere.
    Pflügers Arch.ges.Physiol. 236, 102.
THAUER, R., und G. PETERS (1937). Wärmeregulation nach operativer
    Ausschaltung des Wärmezentrums. Pflügers Arch.ges.Physiol. 239,
    483.
WALTHER, O.-E., M. IRIKI and E. SIMON (1970). Antagonistic changes
    of blood flow and sympathetic activity in different vascular
    beds following central thermal stimulation. II. Cutaneous and
    visceral sympathetic activity during spinal cord heating and
    cooling in anaesthetized rabbits and cats. Pflügers Arch.ges.
    Physiol. 319, 162.
WÜNNENBERG, W., and K. BRÜCK (1970). Studies on the ascending
    pathways from the thermosensitive region of the spinal cord.
    Pflügers Arch.ges.Physiol. 321, 233.
WÜNNENBERG, W., and J.D. HARDY (1971). Responses of units in the
    post. hypothalamus to local temperature changes in the preoptic
    region, the post. hypothalamus, and the spinal cord. Proc.
    Intern. Symposium on Environmental Physiology: Bioenergetics and
    Temperature Regulation. Dublin, July 18 - 23 (in press).

NEURONAL MODELS OF MAMMALIAN TEMPERATURE REGULATION

JOHN BLIGH

Agricultural Research Council Institute of Animal Physiology, Babraham,

Cambridge, England.

AN APPROACH THROUGH MATHEMATICAL MODELS

For so long as it remains impossible to study directly those central nervous complexes which relate the input from temperature sensors to the output to the thermoregulatory effectors, the physiological thermoregulator is like a physical control system sealed in a 'black box' which cannot be opened up for the examination of its components, but the functions of which can be deduced from an analysis of the qualitative and quantitative relations between disturbances and responses. Thus the thermoregulatory functions of the brain, like all other functions of the central nervous system from the simplest spinal reflex to the complex emotional responses to external events, have initially been analysed and described not in the real terms of neural pathways between sensors and effectors, but in terms of the observed relations between disturbances and responses.

Provided these relations are repeatable and sufficiently regular, they might be expressible in terms of mathematics, and mathematical models have been constructed in attempts to define and quantify these relations and to demonstrate their orderliness. However, the central control of body temperature, as with all other integrative activities of the brain, is not an isolated function: the relations between a displacement of skin temperature or core (deep body) temperature and the evoked thermoregulatory responses are dependent on a multitude of independently varying circumstances which presumably influence the neurons which intervene between the input from temperature sensors and output to effectors. These other non-thermal influences may be assumed to vary from moment to moment and to create such complex relations between thermal disturbances and thermoregulatory responses that we cannot reasonably hope to express them in terms of the relatively simple mathematical functions used for the analysis and description of functions with regular linear or curved relations. However, when an organism is kept in a closely controlled environment, the complexity of the interrelating physiological functions is minimised, and the relations between thermal stimuli and thermoregulatory responses can then be measured and expressed as equations. Such mathematical models can be used to predict the thermoregulatory responses of a subject when the experimental conditions are varied beyond those from which the original data used in the construction of the model were obtained (e.g. Atkins and Wyndham, 1969), but because of the complexity of central nervous interactions it might be expected that as the thermal and non-thermal circumstances are moved progressively further away from the restricted conditions used in the construction of the model, the predictive reliability of the model would decline.

Thus a mathematical model is essentially a description of the relations between disturbance and response in very restricted experimental circumstances, and can yield little insight into the nature of the neuronal control responsible for these relations and the resultant control of body temperature. Wyndham and Atkins (1968) have, however, produced a physiological scheme of the control of sweating and heat conductance in man which was derived from the observed relations between peripheral and core temperatures and these thermoregulatory responses. When this scheme is reconstructed in a 'neuronal' format (Fig. 1A) its characteristics are found to be as follows: i) there are two main pathways - one from central warm sensors to heat loss effectors, and the other from central cold sensors to heat production effectors; ii) there are crossed inhibitory connections between these two main pathways, and iii) the neural pathways from peripheral temperature sensors are considered to converge onto these main sensor-effector pathways. In this arrangement it would seem that the activity derived from the peripheral temperature sensors would modify the drive from the primary core temperature sensors to the thermoregulatory effectors.

A

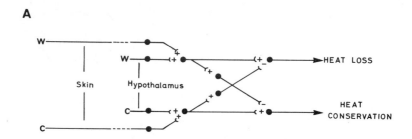

B

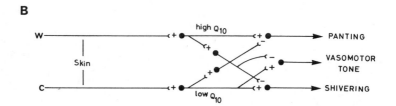

C

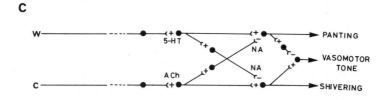

D

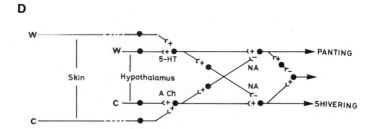

Fig. 1. A. A neuronal network constructed from a scheme of the
            relations between temperature sensors and thermoregu-
            latory effectors suggested by Wyndham and Atkins (1968).
        B. A reconstruction of a neuronal model proposed by Hammel
            (1965).
        C. The neuronal model of Bligh, Cottle and Maskrey (1971).
        D. The neuronal model of Maskrey and Bligh (1972).

AN APPROACH THROUGH ENGINEERING MODELS

Mathematical models have provided a basis for the construction of engineering models which permit descriptions of the controller function in an analogue form. The characteristics of the different types of controller that can be used to regulate temperature in physical systems are well defined and an analysis of the relations between disturbance and response, and of the pattern of the controlled variable (i.e. the temperature which is being regulated) permits deductions to be made about the nature of the control system even when this is inaccessible to direct examination.

Hardy (1961) used this technique to compare the characteristics of mammalian thermoregulation with the regulatory systems most commonly employed, singly or in combination, in industrial and domestic systems of temperature control. His conclusion was that the dominant characteristic of the biological thermoregulator is that of a proportional controller. In such a system there is a reference signal which determines the set point temperature (T_{set}) against which a signal representative of the controlled temperature (T_C) is compared. A derived signal proportional to the differences between T_{set} and T_c is called the load error, and is used to activate the appropriate thermoregulatory responses which are qualitatively related to the direction of the load error and quantitatively related to the magnitude of the load error (Fig. 2).

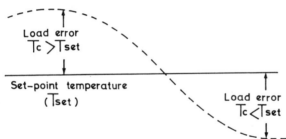

Fig. 2. A diagrammatic representation of the relation between the controlled temperature (T_c) and a reference or set point temperature (T_{set}) in a physical control system.

This analysis has led to the development of a concept of physiological thermoregulation expressed in the analogous terms of a proportional controller in which a signal derived from core temperature sensors, long assumed to be exclusively in the hypothalamic region of the brain, is compared with a stable signal which is unaffected by temperature, which functions as a set point, and which is presumably generated in the thermoregulatory neural complex. This concept is expressed in Fig. 3A. It is supposed that the strength of the signal derived from the hypothalamic temperature sensors is proportional to the local hypothalamic temperature, that the strength of the set point signal is unaffected by changes in hypothalamic temperature, and that the load error is the difference between the strength of these two signals. When the signal from the temperature sensors is greater than that from the set point signal source, this would indicate that core temperature exceeds set point temperature, and a positive load error would give rise to a drive to the heat loss effectors proportional to the extent of the load error. Conversely, when the temperature sensor signal is smaller than the set point signal, this would indicate that core temperature is below the set point temperature and that a negative load error would give rise to a drive to the heat production effectors proportional to the extent of the load

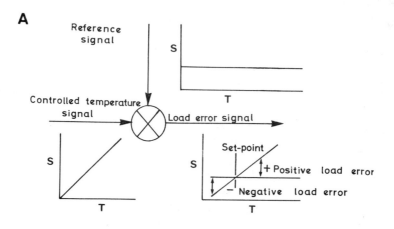

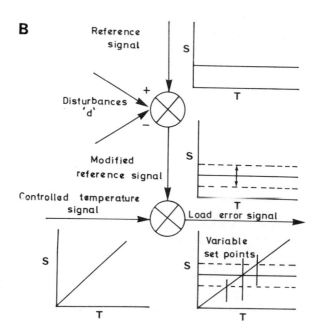

Fig. 3. A. A representation of the relation between the signals from
             a reference source and from a sensor of the controlled
             temperature. The symbol ⊗ represents a signal mixer
          B. A modification of Fig. 3A to indicate that the reference
             signal may be varied by 'disturbance' signals before it
             'mixes' with the controlled-temperature signal.
             S = signal strength; T = temperature.

error.

However, mammalian temperature control clearly does not operate around a fixed set point temperature. In different circumstances core temperature may be shifted to a new equilibrium level. Hammel et al. (1963) suggested that among the various hypothalamic and extrahypothalamic influences which may modify the set point, and thus displace the level at which core temperature is being controlled, are the actions of pyrogens and signals related to factors such as muscular activity and the states of sleep and wakefulness, and afferent signals derived from peripheral temperature sensors. In this way the load error might change and the resultant adjustments in the balance between heat production and heat loss would lead, secondarily, to a shift in core temperature. In other words, a displacement of core temperature is not necessarily a primary stimulus.

The variability of the set point temperature (T_{set}) is expressed in Fig. 3B in which the sum of the disturbances (d) to the set point signal modifies the temperature set point to $T_{set} \cdot d$, and the load error which supplied the thermoregulatory drive to $T_{set} \cdot d - T_c$. Thus the load error may vary because of a change in $T_{set} \cdot d$ and in the absence of a prior change in T_c. However, because of the resultant change in the balance between heat production and heat loss a change in the level of core temperature might then occur if a major component of 'd' continues to exert an influence on the set point temperature.

The model discussed above is a purely analogue description of the physiological thermoregulator, and does not bring us very much nearer to an understanding of the neurophysiology of thermoregulation unless we can form a concept of how functions comparable to these physical functions might be achieved by synaptic interactions between neurons.

THE INTERPRETATION OF THE ELECTRICAL ACTIVITIES OF SINGLE HYPOTHALAMIC NEURONS

During the past ten years studies have been made in several laboratories, and on several species, of the relations between local hypothalamic temperature and the electrical activities of single hypothalamic neurons. These have yielded evidence of many different patterns of neuronal activity which have been recently collated by Guieu and Hardy (1971). Five of these many different patterns, which will be referred to later in this essay, are illustrated diagrammatically in Fig. 4. Those units the activity of which varied positively and linearly with local temperature (Fig. 4A) are considered to be primary warm sensors. Those units which showed a negative linear relation between activity and temperature are considered to be primary cold sensors. Other neurons the activity of which was altered by local changes in temperature but which yielded a variety of biphasic relations between activity and temperature (e.g. Fig. 4 C,D) are considered to be interneurons which are indirectly influenced by local temperature changes acting on the primary temperature sensors. Neurons which are almost completely temperature insensitive (Fig. 4E) are considered to be concerned in the set point function.

At present it is not clear whether the several types of biphasic patterns which have been reported are characteristic of specific neurons each having a definite though as yet undetermined function in the hypothalamic pathways between temperature sensors and thermoregulatory effectors. Some of these different biphasic patterns may relate only to transient circumstances which were prevailing at the time of recording, and which influenced the sum of the excitatory and inhibitory influences acting synaptically on the interneurons. Possibly many different activity/temperature patterns could be obtained from the same interneuron in different circumstances.

Despite this uncertainty, Hardy and Guieu (1971) have assumed that these different activity/temperature patterns are characteristic of particular neurons, and they have utilised most of the patterns which have been obtained in different species, and some of which have been encountered only once, to construct a complex neuronal model. The ingenuity of this assembly is considerable, but its complexity may be a handicap rather than a help at the present stage in our attempts to consider thermoregulation in terms of neurophysiology. When the complexity of a model approaches that of reality, it can become as difficult to

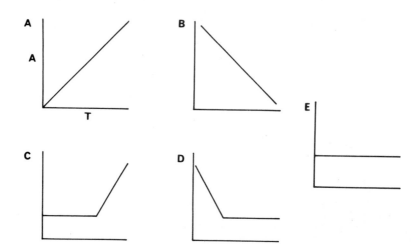

Fig. 4.   Diagrammatic representations of five distinct patterns selected
          arbitrarily from the many reported patterns of the relations
          between the electrical activity of single preoptic/anterior
          hypothalamic neurons and local hypothalamic temperature.
          A and B are considered to be the characteristics of primary warm-
          and cold-sensors respectively.
          C and D are considered to be characteristics of interneurons on
          the pathways between warm- and cold-sensors and the heat loss and
          heat production thermoregulatory effectors respectively.
          E is a characteristic of temperature-insensitive neurons which
          are considered to be of thermoregulatory significance (see text).

understand and to test, and therefore loses much of its usefulness. It is essential, at this stage, that we discuss neuronal patterns in the simplest possible
terms, and only build them up into more complex representations as the evidence
necessitates.

        The first speculative attempt to consider how the control of body temperature
might be achieved neuronally was that of Bazett (1949) who suggested that hypothalamic warm sensitive and cold sensitive neurons might have different activity/
temperature characteristics. Vendrik (1959) has interpreted Bazett's proposal in
terms of the discharge frequencies of temperature sensors which rise to a peak
value and then decline as temperature rises (Fig. 5A). If the temperature for
peak activity of the cold sensors is below the normal controlled level of core
temperature, and that of the warm sensors is above this level, then when core temperature rises above the point where the activity of cold- and warm-sensors coin-
cide, the activity of the one will decline and that of the other will increase. On
the assumption that the cold sensors are linked in some way to the outflow to the
heat production effectors, and that the warm sensors are linked to the outflow to
the heat loss effectors, the effect of an upward displacement of core temperature
will be to decrease heat production and to increase heat loss. This adjustment in
the balance between heat loss and heat production will tend to drive core temperature back to the point of balance. Similarly, if core temperature falls below
this point of balance, the activity of the cold sensors and of heat production
effectors will be increased, while that of warm sensors and of heat loss effectors
will be reduced, and core temperature will again be driven back to the point of
balance. Thus Bazett's idea, as interpreted by Vendrik (1959), was that the set
point is determined by the different characteristics of these two populations of
primary temperature sensors.

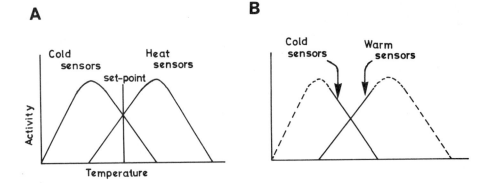

Fig. 5. A. A diagrammatic representation of the activity/temperature
             characteristics of hypothalamic cold- and warm-sensors
             suggested by Vendrik (1959).
         B. The positive and negative linear activity/temperature re-
             lations of hypothalamic neurons considered to be primary
             warm- and cold-sensors (Fig. 4A and B) shown as segments
             of the bell-shaped characteristics of these temperature-
             sensors suggested by Vendrik (1959).

    This entirely theoretical but original proposal is of considerably more than
historical interest because subsequent studies of the electrical activities of
single hypothalamic neurons indicated that when temperature is varied within the
normal range of core temperature, some units show a positive linear relation between activity and temperature, while others show a negative linear relation, over
this range (Fig. 4 A,B). Obviously this linearity cannot persist over a wide
temperature range: there must be a peak of activity and subsequent decline. Thus
these linear relations might be regarded as belonging to the effective portions of
activity/temperature curves similar to those proposed by Vendrik (Fig. 5B), and
indeed, when Cabanac et al. (1968) plotted activity/temperature relations of temperature-sensitive hypothalamic neurons over a wider range of hypothalamic temperature they found the predicted bell-shaped curves. These curves did not fall into
two entirely distinct populations but nevertheless the peaks of activity of some
neurons occurred at hypothalamic temperatures above normal core temperature, while
those of others occurred at temperatures below normal core temperature.
    If it is supposed that the primary hypothalamic temperature sensors in the
reconstructed model of Wyndham and Atkins (1968) (Fig. 1A) are neurons with bell-
shaped characteristics similar to those proposed by Vendrik (1959) and for which
Cabanac et al. (1968) found evidence, then it might also be supposed that the
basic set point mechanism of mammalian thermoregulation depends on these characteristics, with the overall balance between heat production and heat loss at any
given level of hypothalamic temperature affected by the synaptic influence of
peripheral thermosensitivity and, perhaps, by various other non-thermal excitatory and inhibitory influences.
    However, the temptation to regard the Bazett/Vendrik hypothesis as vindica-
ted and to attribute the set point of body temperature to the characteristics of
primary cold sensors and warm sensors in the hypothalamus is tempered by a persistent uncertainty whether cold sensors really exist in this region of the brain.
If they do, they would seem to be much less numerous than the warm sensors.
    In one of the earliest studies of the electrical activity of single neurons
in the preoptic/anterior hypothalamus, Nakayama et al. (1963) found that a minor-
ity of neurons responded to a local rise in body temperature with increased
activity, but that the activity of the majority of the neurons was unaffected by
local changes in temperature. Hammel (1965) described the temperature-sensitive

neurons as 'high Q_{10} units', and the temperature-insensitive neurons as 'low Q_{10} units', and proposed that the set point depends on the interaction of these two populations of neurons. This concept is expressed in Fig. 6A. The set point temperature is considered to be that temperature at which the activities of the two populations of neurons are equal. Above this temperature the activity of the high Q_{10} neurons is greater than that of the low Q_{10} neurons, and the thermoregulatory responses are those to be expected when core temperature is above a set point temperature: the drive to heat loss effectors is increased, and that to heat production effectors is reduced. The converse situation exists when core temperature is below the set point temperature. There is an apparent relation between this neuronal concept and that derived from an engineering analogue (Fig. 3A). Although Hammel (1965) did not mention the earlier proposal of Vendrik (1959) it is also apparent that these two neuronal concepts are very similar.

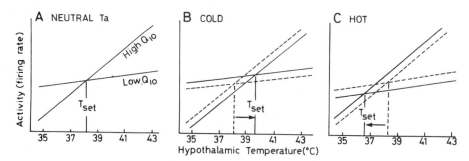

Fig. 6.   The relations between the activity/temperature characteristics
          of the high Q_{10} and the low Q_{10} neurons as proposed by Hammel
          (1965). The set point of temperature regulation is considered
          to be at that temperature at which the activities of the two
          populations of neurons are the same. A illustrates the set
          point temperature at a thermoneutral temperature. In a cold
          environment (B) the change in the activity/temperature charac-
          teristics of the neurons results in an upward shift in the set
          point temperature. In a hot environment (C) there is a down-
          ward shift in the set point temperature.

As has been mentioned above, the level at which body temperature is apparent-ly regulated is affected by a number of physiological circumstances including changes in the temperature of the skin. This adjustment in the set point tempera-ture in response to changes in skin temperature was attributed by Hammel (1965) to a shift in the activity/temperature characteristics of both the high Q_{10} and the low Q_{10} neurons, so that the local hypothalamic temperature at which the activit-ies of the two populations of neurons were coincident was raised in response to a fall in skin temperature (Fig. 6B), and lowered in response to a rise in skin tem-perature (Fig. 6C). These shifts in set point temperature could account for an immediate increase in heat production in response to a fall in skin temperature, and an immediate increase in heat loss in response to a rise in skin temperature in the absence of any prior change in hypothalamic temperature. A neuronal arrangement to account for this influence of skin temperature on thermoregulatory responses and on the level of core temperature was proposed by Hammel (1965). When this model is reconstructed so that it can be readily compared with the re-constructed scheme of Wyndham and Atkins (1968) (Fig. 1B), it is evident that the-re are both similarities and differences between the two neuronal models. The similarities are: i) the two main neural pathways from hypothalamic temperature sensors to thermoregulatory effectors, ii) the crossed inhibitory connections between these two main pathways, and iii) the indication that the neural pathways from the peripheral temperature sensors impinge on these main pathways and act to

modify the drives to the thermoregulatory effectors at any given level of hypothalamic temperature. The major difference between these two models is that
Hammel suggests that the high Q_{10} and low Q_{10} neurons are interneurons upon which
the pathways from the peripheral temperature sensors impinge.

In the construction of a purely theoretical neuronal model, there is little
to choose between that based on the scheme of Wyndham and Atkins (1968) and that
proposed by Hammel (1965). Whether the pathways from the peripheral temperature
sensors act directly on the central temperature sensors, or converge at a subsequent synapse, is not crucial to the concept of a variable set point. In both
models, the change in the input from the peripheral temperature sensors would
change the drive from hypothalamic sensors to the thermoregulatory effectors, and
would thus change the balance between heat production and heat loss. Also,
whether the hypothalamic warm sensors and cold sensors are high Q_{10} and low Q_{10}
neurons, or have positive and negative linear temperature characteristics is not
fundamental to the basic concept of how the physiological temperature regulator
may be functioning.

A MODEL DERIVED FROM SYNAPTIC INTERFERENCE STUDIES

Some of the putative transmitter substances which occur naturally in the
brain have distinct and reproducible effects on thermoregulatory effector functions and on core temperature when they are introduced, in pharmacological quanti-
ties, into the cerebral ventricles or directly into the preoptic/anterior hypothalamic (PO/AH) region of the brain.

The implications and limitations of these studies are discussed in the accom-
panying essay by Dr Richard Hellon. Here, the discussion of such synaptic interference studies is confined to a representation of the neuronal models of Bligh
et al. (1971) and of Maskrey and Bligh (1972) derived from studies of the thermoregulatory effects of injections into the cerebral ventricles of sheep of 5-
hydroxytryptamine (5-HT), noradrenaline (NA), and acetylcholine, eserine and
acetyl-choline, or carbachol (all referred to by the abbreviation ACh), at different ambient temperatures, or when the anterior hypothalamus is locally heated or
cooled.

The interactions between ambient temperature and the intraventricularly injected putative transmitter substances were precisely as if the substances were
acting, as indicated, at synapses with a neuronal arrangement given in Fig. 1C.
5-HT had little or no effect on thermoregulation when ambient temperature was high
and when, it is presumed, the activity of the warm sensor to heat loss pathway
was already maximally activated, and the cold sensor to heat production pathway
was already maximally inhibited. At the low ambient temperature, however, 5-HT
caused an increase in respiratory frequency, an inhibition of shivering, and a
resultant fall in core temperature. Conversely, cholinomimetic substances (ACh)
were relatively ineffective at low ambient temperatures when shivering was already
present and respiratory frequency was depressed. At high ambient temperatures
these substances caused the depression of panting and/or the initiation of shivering, and a rise in core temperature. Noradrenaline was found to inhibit all
thermoregulatory effector processes: when it was injected at a high ambient temperature, panting was inhibited and core temperature rose; when it was injected at
low ambient temperature, shivering was inhibited and core temperature fell. Changes in peripheral vasomotor tone consistent with the model were seen only when
ambient temperature was close to the range of thermoneutrality for the sheep. At
high ambient temperatures the peripheral vessels remained dilated, and at low
ambient temperatures, the vessels remained constricted.

In these studies by Bligh et al. (1971) the location of the temperature sensors was undefined, but it may be assumed that the principal effect of ambient
temperature was on the peripheral temperature sensors. In the subsequent experiments of Maskrey and Bligh (1972) the ambient temperature was held constant and
the hypothalamic temperature was locally raised or lowered while the substances
were injected into the cerebral ventricles. The results were expressable in terms
of the same neuronal model (Fig. 1C). The similarity of the effects of changes in
peripheral and hypothalamic temperatures on the thermoregulatory effects of cen-

trally administered putative synaptic transmitter substances implies the convergence of the pathways from peripheral and central temperature sensors. This implication is represented in the modified neuronal model (Fig. 1D) of Maskrey and
Bligh (1972).

The model is not intended to imply that 5-HT and ACh are necessarily the
transmitter substances liberated by the primary temperature sensors: there could
be intervening neurons between the convergence of the pathways from peripheral and
central temperature sensors and the synapses where the injected substances seem to
be acting. Nor is this model intended to imply a similar pattern of pathways and
synapses in the thermoregulatory neuron pools of other species. That the model
does not have general validity is clear from the ample evidence of quite different
effects of these putative transmitter substances on the central regulation of
temperature of other mammalian species.

It is noteworthy that that basic model to describe our experiments on sheep
(Fig. 1C) had been formulated before I had converted the models of Hammel (Fig. 1B)
and of Wyndham and Atkins (Fig. 1A) into comparable formats. Thus the similarity
of these three models is all the more striking because each was derived from quite
different kinds of evidence: Hammel's model was largely determined by the relations between the activity and temperature of single neurons in the anterior hypothalamus, although it may also have been influenced by Hammel's earlier consideration of engineering models; Wyndham and Atkins' model was determined largely by
a study on man of the thermoregulatory responses to variations in skin and core
temperatures. Our model was an attempt to give expression to the interactions in
the sheep between the effects of different levels of either ambient or hypothalamic temperature and the effects of substances injected into the cerebral ventricles
on thermoregulation.

UNIT ACTIVITY EVIDENCE OF CONVERGENT PATHWAYS FROM TEMPERATURE SENSORS

While those hypothalamic neurons the activity of which responds linearly with
local temperature changes are generally considered to be primary temperature sensors (i.e. to be directly temperature sensitive), there is no reliable and undis-
puted evidence that these units are primary neurons which are quite unaffected by
other neurons. Unit activity studies have yielded clear evidence of a convergence
of the pathways from peripheral, spinal and hypothalamic temperature sensors, but
it is not clear exactly at what stage in the pathways this occurs. The cutaneous,
spinal and hypothalamic temperature sensors may all be primary neurons which converge onto a common path, or the hypothalamic temperature sensors might be interneurons onto which the pathways from the other temperature sensors impinge. Wit
and Wang (1968) found that the activities of some of the hypothalamic units from
which they recorded were influenced by both local temperature changes and skin
temperature changes, while other units were influenced only by hypothalamic temperature changes. This might imply the convergence suggested in the models of
Wyndham and Atkins (1968) and Maskrey and Bligh (1972). Hellon (1969), by contrast, found that all the cells in the PO/AH region of the rabbit which were
affected by changes in local temperature were also affected by changes in skin
temperature. This might imply the convergence suggested in the model of Hammel
(1965). However, in a more recent and more detailed report, Hellon (1970) indicates that only some of the PO/AH neurons in the rabbit which are affected by local
temperature changes are also affected by changes in skin temperature. A personal
preference for the arrangement in Fig. 1D cannot be defended except as a matter
of convenience in presenting and discussing neuronal models.

Evidence that in the dog (Jessen et al., 1967; Jessen, 1967) and in the ox
(Hales and Jessen, 1969) local heating of the spinal cord can give rise to thermoregulatory patterns very similar to those which result from cutaneous and hypothalamic heating indicates that there may be a convergence of pathways from temperature sensors from many locations in the body including the spinal cord. Cen-
tripetal pathways from the spinal temperature sensors were demonstrated by Simon
and Iriki (1970), and Guieu and Hardy (1969) recorded from a neuron in the PO/AH
region of the rabbit, which was classed as an interneuron on the basis of its
activity/temperature relations, and which was inhibited by local heating of both

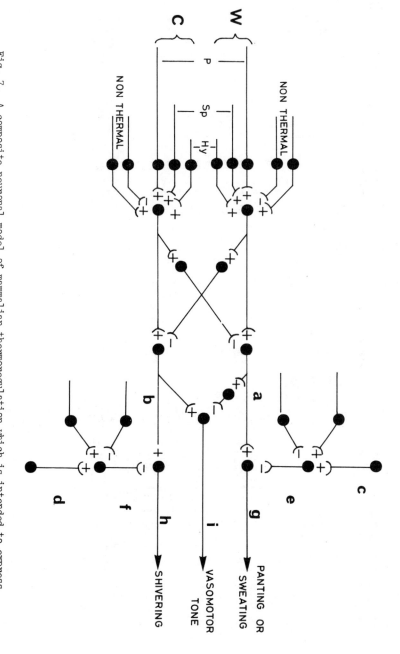

Fig. 7. A composite neuronal model of mammalian thermoregulation which is intended to express the variable and complex nature of the input to the pathways between temperature sensors and thermoregulatory effectors, and the variable threshold temperatures for heat production and evaporative heat loss. See text for details. The letters a to i relate features in Fig 7 and 8 to each other and to the text. W = warm sensors; C = cold sensors; P = peripheral; Sp = spinal cord; Hy = hypothalamus.

the PO/AH region and the spinal cord. This convergence of pathways from temperature sensors in the hypothalamus, spinal cord and the skin is expressed in a modification of the model by Maskrey and Bligh (Fig. 7). The convergence is not necessarily simple: some evidence of Nakayama and Hardy (1969) suggests that the
pathways from extrahypothalamic temperature sensors may converge before they reach
the PO/AH region. However, at the present stage in the development of neuronal
models of thermoregulation, this is a detail which may not be important to the
basic concepts.

AN INTEGRATIVE AND SPECULATIVE NEURONAL MODEL

Fig. 7 shows afferent pathways from both cold sensors and warm sensors in the
three demonstrated regions of temperature sensitivity. An interrelation between
the thermoregulatory effects of spinal heating and cooling and the injections of
5-HT, ACh and NA into the cerebral ventricular system of sheep has not been investigated, and the sensory input in Fig. 7 is a synthetic concept based on both
the unit activity studies of other workers and our own synaptic interference
studies. Whether cold sensors and warm sensors occur at each location, whether
additional thermal information is derived from other tissues, and whether species
vary greatly in the distribution and relative concentrations of cold sensors and
warm sensors, remain open questions.
Snellen (1966) has reported that the sweat rate of man can be best correlated
not with core temperature nor with skin temperature, but with mean body temperature. Somewhat similarly, Brück and Wünnenberg (1967) have reported that shivering in the guinea-pig can also be best correlated with mean body temperature.
While these results may mean that there is a convergence of information from hypothalamic and extra-hypothalamic core temperature sensors and from peripheral
temperature sensors, it does not mean that the contributions from these different
temperature sensitivity compartments are necessarily equal. This would, indeed,
be unlikely, and in different circumstances and in different species, the temperature with which thermoregulatory effector functions can be best correlated may be
nearer to that of the core or of the skin. In other words, mean body temperature
should be regarded as an approximate correlate rather than a definite and fixed
correlate.
The convergence of neural pathways onto the main pathways to the thermoregulatory effectors in Fig. 7 includes two additional inputs to each main pathway, one of which is excitatory and the other inhibitory. These are intended to
represent all the events in the central nervous system related to non-thermal
disturbances which may influence the relations between temperature sensors and
thermoregulatory effectors.
If the convergence were, directly or indirectly, as represented in Fig. 7,
then the intensity of the drive to the thermoregulatory effectors would depend on
the summated influence of the information from all the temperature sensors where-
ever they are located, together with that of all the extrahypothalamic and non-
thermal influences which impinge upon the neuronal pools concerned in temperature
regulation. The degree of influence of peripheral and of central temperature
sensors, which was once discussed in absolute terms, probably varies with conditions. Experimentally, an adequate displacement of skin-, spinal- or hypothalamic-
temperature might give rise to a sufficiently strong synaptic influence to domin-
ate the drive to the thermoregulatory effectors, but such experiments may give
little indication of the relative influences of these different categories of
temperature sensors in more natural circumstances when the whole body temperature
is being displaced much more slowly.
On the not unlikely assumption that, for the larger mammals at least, core
temperature sensors (hypothalamic, spinal and others) are numerically dominant
over peripheral temperature sensors, and that the core temperature sensors consist of two populations as proposed by Bazett (1949) or Hammel (1965), then the
essential stability of core temperature could reside in the properties of these
core temperature sensors. Variations in the level of core temperature could be
attributed to the synaptic influences of peripheral temperature sensors and of
the non-thermal influences.

The composite afferent input in Fig. 7 might go some way towards a representation of various thermal and non-thermal influences on the balance between heat production and heat loss, and therefore on core temperature under steady state conditions, but it fails to account for the different threshold temperatures for the activation of evaporative heat loss by panting or sweating, and of heat production by shivering, and the resultant time delays between the thermal stimulus and these thermoregulatory responses. That the thresholds for evaporative heat loss and for heat production are dependent on both skin and core temperatures has been demonstrated by Chatonnet et al. (1964) and by Cabanac et al. (1965). These relations can be explained in terms of the afferent input in the neuronal model (Fig. 7) if the contributions of peripheral and core temperature sensors to common neural pathways are proportional to the degree to which they are activated, but the integration of input signals cannot account for the delay in the onset of sweating or panting and of shivering until a threshold core temperature has been reached.

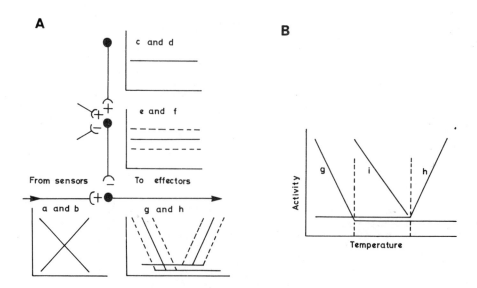

Fig. 8. A.   A schematic representation of the suggested role of the temperature insensitive neurons as the source of inhibitory influences which act on both the warm sensors to evaporative heat loss effectors, and the cold sensors to heat production effector pathways. This inhibition is considered to block transmission from sensors to effectors until the excitatory activity of the sensors exceeds this inhibitory influence.

If the basic level at which body temperature is regulated should depend on the positive and negative linear activity/temperature characteristics of two populations of primary temperature sensors (Fig. 4A,B), then the functions of the temperature insensitive neurons (Fig. 4E) remains obscure. A possible function for them can be derived from the engineering concepts summarised in Figs 3 A and 3B, and from the five selected activity/temperature patterns in Fig. 4. The temperature insensitive neurons might give rise to a constant or variable (Fig. 8A) inhibitory influence which acts on each of the main neural pathways between temperature sensors and thermoregulatory effectors. In Fig. 7 these inhibitory influences are placed beyond the points from which the influences on peripheral

vasomotor tone stem. Thus placed, the effect of these inhibitory influences
could be to block synaptic transmission to evaporative heat loss and heat production effectors until the intensity of the drive along the main pathways from the
sensors becomes sufficient to overcome the inhibition, without impeding the
control of peripheral vasomotor tone. The physiological consequence of this neuronal network and pattern of electrical activities might be as shown in Fig. 8B.
The drive to the heat production effectors (g) is initiated at a variable lower
threshold core temperature, and its intensity then increases linearly to a maximum
as core temperature falls further. The drive to evaporative heat loss (h) is
initiated at a variable upper threshold core temperature, and its intensity increases linearly to a maximum as core temperature rises further. The drive to
peripheral vasomotor tone is considered to vary linearly in the 'neutral' range
between these two threshold temperatures. The drive is supposed to be maximal at
the threshold temperature for shivering, so the cutaneous vessels are then maximally constricted, and to be minimal at the threshold temperatures for evaporative heat loss.

The definition of these threshold temperatures in terms of core temperature
may be related more to orthodoxy than to physiological reality, but these thresholds for the different thermoregulatory effector functions need to be expressed
in terms of an identifiable and measurable body temperature. For this purpose a
core temperature is an obvious choice. However, in terms of this model (Fig. 7),
the variabilities of the set point and threshold temperatures relate to this
arbitrary choice of a reference body temperature. This may simply be a means of
expressing changes in the balance between heat production and heat loss consequent
on a change in composition of afferent influences acting on the two main neural
pathways through the thermoregulatory 'centres'.

The development of the neuronal model in Fig. 7 is, of course, simply a
rationalization of the functions of selected unit activity patterns (Fig. 4)
which may have something to do with temperature regulation. Theoretically, the
scheme is attractive in that it provides for a continuous control of peripheral
vasomotor tone depending on the relative intensities of the summated drives along
the two main pathways from temperature sensors to thermoregulatory effectors,
while the thresholds for the onset of panting and/or sweating and shivering depend
on the build-up of sufficient temperature sensor activity to overcome the inhibitory influences. The 'thresholds' for these thermoregulatory functions are
traditionally expressed in terms of core temperature, and the variability of the
core temperature at which threshold is reached may thus depend on the intensity
of other thermal and non-thermal contributions to the activity along these main
neural pathways. The 'thresholds' may also depend on the sum of the excitatory
and inhibitory drives acting synaptically to modify the intensity of the inhibitory influences derived from the temperature-insensitive neurons. (Fig. 8A).

A question which might be asked is whether, in the terms of this theoretical
neuronal model, the level of core temperature is basically a consequence of the
activity/temperature characteristics of the temperature sensors, or of the threshold temperatures for evaporative heat loss and heat production created by the
inhibitory influences, or to a combination of the characteristics of the sensors
and the inhibitory influences. A recent chance observation in my laboratory
(Maskrey and Bligh, unpublished) indicates that the so-called set point may be
independent of the integrity of the pathways between temperature sensors and
thermoregulatory effectors.

A sheep into which we had implanted a multiple hypothalamic thermode assembly
and in which we had cannulated a lateral cerebral ventricle (Maskrey and Bligh,
1972) was unable to restore its core temperature to its normal level after this
had been raised or lowered by the thermoregulatory responses to heating or cooling of the skin or the anterior hypothalamus, or to intraventricular injections
of 5-HT, NA and ACh. These thermoregulatory responses were all quite normal.
Thus the syndrome was of an animal in which the neural connections between temperature sensors and thermoregulatory effectors were unimpaired but in which the set
point mechanism had been inactivated. This observation indicates that the set
point involves neural structures which are collateral to the main sensor to
effector pathways.

Since the effect of a pyrogen on body temperature is said to be due to an elevation of the set point at which body temperature is regulated, the above interpretation was strengthened by the further observation that this sheep did not develop a fever in response to an intravenous injection of TAB-vaccine in a dose which would produce a fever in the sheep in normal circumstances.

CONCLUDING COMMENT

As the neuronal model derived from our own synaptic interference studies evolved, we were aware of the danger of confusing it with reality. - It is only what any working model should be: a simple illustration of all the observations. Our recent efforts to find the limitations of the model as a description of experimental observations have, so far, been unsuccessful but we are sure that it is only a matter of time before the model will have to be modified beyond recognition, or totally superceded.

The extensions which I have built onto this model during the construction of this essay must be viewed with the same degree of scepticism. It is only a model, and it has only two functions: i) to summarize in the simplest possible terms, the observed patterns in the relations between thermal disturbances, thermoregulatory responses and the level of core temperature, and ii) to present an interpretation of how existing evidence, placed in a neuronal format, might suggest experiments from which the truth will ultimately emerge.

REFERENCES

Atkins, A.R. and C.H. Wyndham, (1969), A study of temperature regulation in the human body with the aid of an analogue computer, Pflügers Arch. ges. Physiol. 307, 104.

Bazett, H.C. (1949), The regulation of body temperatures In: Physiology of heat regulation and the science of clothing. Ed. L.H. Newburgh (Philadelphia: Saunders).

Bligh, J., W.H. Cottle and M. Maskrey, (1971), Influence of ambient temperature on the thermoregulatory responses to 5-hydroxytryptamine, noradrenaline and acetylcholine injected into the lateral cerebral ventricles of sheep, goats and rabbits, J. Physiol., Lond. 212, 377.

Brück, K. and W. Wünnenberg, (1967), Die Steuerung des Kältezitterns beim Meerschweinchen, Pflügers Arch. ges. Physiol. 293, 215.

Cabanac, M., J. Chatonnet and R. Philipot, (1965), Les conditions de temperatures cerebrale et cutanee moyennes pour l'apparition du frisson thermique chez le Chien, C. r. hebd. Seanc. Acad. Sci., Paris, 260, 680.

Cabanac, M., J.A.J. Stolwijk and J.D. Hardy, (1968), Effect of temperature and pyrogens on single-unit activity in the rabbit's brain stem, J. appl. Physiol. 24, 645.

Chatonnet, J., M. Cabanac and M. Mottaz, (1964), Les conditions de temperatures cerebrale et cutanee moyenne pour l'apparition de la polypnee thermique chez le Chien, C.R. Seanc. Soc. Biol. 158, 1354.

Guieu, J.D. and J.B. Hardy, (1969), Changes in single unit activity in the pre-optic region due to temperature changes in the spinal cord, Physiologist. 12, 243.

Guieu, J.D. and J.D. Hardy, (1971), Integrative activity of preoptic units I: response to local and peripheral temperature changes, J. Physiol., Paris. 63, 253.

Hales, J.R.S. and C. Jessen, (1969), Increase of cutaneous moisture loss caused by local heating of the spinal cord in the ox, J. Physiol., Lond. 204, 40P.

Hammel, H.T. (1965), Neurones and temperature regulation In: Physiological regulation and control, Eds. W.S. Yamamoto and J.R. Brobeck (Philadelphia: Saunders).

Hammel, H.T., D.C. Jackson, J.A.J. Stolwijk, J.D. Hardy and S.B. Strømme, (1963), Temperature regulation by hypothalamic proportional control with an adjustable set-point, J. appl. Physiol. 18, 1146.

Hardy, J.D., (1961), Physiology of temperature regulation, Physiol. Rev. 41, 521.

Hardy, J.D. and J.D. Guieu, (1971), Integrative activity of preoptic units
    II: hypothetical network, J. Physiol., Paris. 63, 264.

Hellon, R.F., (1969), Environmental temperature and firing rate of hypothalamic
    neurones, Experientia. 25, 610.

Hellon, R.F., (1970), The stimulation of hypothalamic neurones by changes in
    ambient temperature, Pflügers Arch. ges. Physiol. 321, 56.

Jessen, C., (1967), Auslösung von Hecheln durch isolierte Wärmung des Rüchen-
    marks am wachen Hund, Pflügers Arch. ges. Physiol. 297, 53.

Jessen, C., K-A. Meurer and E. Simon, (1967), Steigerung der Hautdurchblutung
    durch isolierte Wärmung des Ruchenmarks am wachen Hund, Pflügers Arch. ges.
    Physiol. 297, 35.

Maskrey, M. and J. Bligh, (1972), Interactions between the thermoregulatory res-
    ponses to injections into a lateral cerebral ventricle of the Welsh Mountain
    sheep of putative neurotransmitter substances, and of local changes in
    anterior hypothalamic temperature, Int. J. Biometeorol. (in press)

Nakayama, T., H.T. Hammel, J.D. Hardy and J.S. Eisenman, (1963), Thermal stim-
    ulation of electrical activity of single units of the preoptic region,
    Am. J. Physiol. 204, 1122.

Nakayama, T. and J.D. Hardy, (1969), Unit responses in the rabbit's brain to
    changes in brain and cutaneous temperature, J. appl. Physiol. 27, 848.

Simon, E. and M. Iriki, (1970), Ascending neurons of the spinal cord activated
    by cold, Experientia. 26, 620.

Snellen, J.W., (1966), Mean body temperature and the control of thermal sweating,
    Acta Physiol. pharmac. neerl. 14, 99.

Vendrik, A.J.H., (1959), The regulation of body temperature in man, Ned. T.
    Geneesk. 103, I, 1.

Wit, A. and S.C. Wang, (1968), Temperature-sensitive neurons in preoptic/anterior
    hypothalamic region: effects of increasing ambient temperature, Am. J.
    Physiol. 215, 1151.

Wyndham, C.H. and A.R. Atkins, (1968), A physiological scheme and mathematical
    model of temperature regulation in man, Pflügers Arch. ges. Physiol.
    303, 14.

THE SET-POINT IN TEMPERATURE REGULATION: ANALOGY OR REALITY

H. T. HAMMEL
Physiological Research Laboratory
Scripps Institution of Oceanography
University of California, San Diego
La Jolla, California, USA

A temperature regulator may be defined as an organism that can
activate appropriate behavioral and autonomic responses which
minimize deviations of the core temperature from an optimal temperature for some or all of the body functions. This definition does
not specify the precision of the regulator; it does not specify the
optimal temperature nor does it specify that regulation occur at all
times during the life of the organism. The definition does imply
that there exists some mechanism for comparing body temperature with
optimal temperature. If the difference is plus, i.e. body temperature greater than optimal temperature, at least one response can
be made to reduce the body temperature. Likewise, if the difference
is minus, at least one response can be activated to increase the
body temperature. I suppose if we knew how the body senses its
temperature and compares it with the optimal temperature, we would
be well along toward an answer to the query: is the set-point in
temperature regulation an analogy or a reality? I think we are
seeking a neural network or a flow chart involving neurons which
incorporates the properties for sensing temperature and analyzing
this with respect to the optimal temperature and other information.

SOME IMPORTANT CHARACTERISTICS OF THE TEMPERATURE REGULATING NETWORK

First, I would like to review some of the known properties of
the peripheral and central nervous system which I think are important
and must be incorporated into a neural network which purports to
describe temperature regulation. These properties are:
  1) the dependence of the cutaneous temperature transducers
     upon temperature - both the steady state and phasic
     dependence,
  2) the dependence of thermoregulatory responses upon preoptic-
     anterior hypothalamic (POAH) temperature in the resting
     animal,
  3) the negative feedback resulting from prolonged displacement
     of the POAH temperature,
  4) the effect of exercise upon the relationship between POAH
     temperature and thermoregulatory responses,
  5) the effects of the state of hibernation upon the relation-
     ship between POAH temperature and thermoregulatory
     responses.

I would like to proceed by illustrating the first three properties
and suggest a basic neural network purporting to possess these
properties. Then I shall return to and illustrate the next two
properties and modify the neural network accordingly.

Some characteristics of cutaneous temperature receptors

   Common experience and neurophysiological evidence are convincing
that some property of temperature is transduced to neural impulses
in afferent fibers from the skin of furred animals. In fact, two
classes of fibers are found from furred skin, "cold" and "warm"
fibers, Fig. 1. The bell shaped curve on the left was obtained
from the average of adapted firing rates from several "cold" fibers
in the cat skin as a function of steady state temperature (Iggo,
1970).

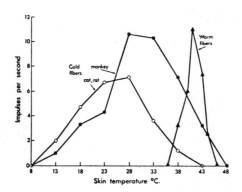

     Fig.   1.   Sensitivity curves for cutaneous thermoreceptors.
                 Each point on the graph is the mean adapted
                 frequency of discharge for several units. The
                 "cold" receptors in the cat and rat have a peak
                 sensitivity at a lower temperature than the
                 monkey; these units are excited by a fall in
                 skin temperature. The "warm" receptors, on the
                 other hand, are excited by a rise in temperature.
                 (Based on H. Hensel, A. Iggo and I. Witt, J.
                 Physiol., 153: 113-126, 1960; Iggo, J. Physiol.,
                 200:403-430, 1969.) Reproduced by permission
                 (Iggo, 1970).

These fibers were nonmyelinated and showed peak activity in the
middle 20's°C. Monkey skin has a similar class of "cold" fibers
which are myelinated and faster conducting and their peak activity
was in the low 30's°C. "Cold" fibers from these animals, as well as
from rat, dog, rabbit and man, show steady activity at skin temperatures prevailing in a thermally neutral environment; and in fact,
their adapted activity goes to zero only for skin temperatures well
above internal body temperature. These fibers also show a phasic
response to a changing skin temperature; the firing rates for a
falling temperature are considerably greater than the adapted rates
at the same temperature and the firing rates trend to zero for a
rising skin temperature.

   The steady state curve for "warm" fibers is shown on the right.
"Warm" fibers show little or no activity at skin temperatures
prevailing in a neutral environment and show peak activity at
temperatures well above internal temperature and at levels that would

seldom if ever occur at the skin surface. However, "warm" fibers
also exhibit phasic activity, i.e. they are activated by a rising
skin temperature at temperatures well below the threshold steady
state temperature shown here. These, then, are some important
properties of the cutaneous temperature transducers.

Some characteristics of the POAH nuclei

Now we turn to another temperature transducing part of the
nervous system, the preoptic-anterior hypothalamic nuclei which
lie between the optic chiasm and the anterior commissure. This
tissue was investigated by surrounding it with thermodes which could
be perfused with water so as to fix the temperature of the tissue
at any level between 31 and 42°C. Any of several thermoregulatory
responses can be measured while varying the POAH temperature over
this range in the resting or running animal exposed to a range of
environmental temperatures.

For example, the evaporative heat loss from the respiratory tract
of a resting dog was a function of the POAH temperature, as shown in
Fig. 2 for four environmental temperatures. In the 35°C environment,
upper left, evaporative heat loss remained at the insensible level
for all POAH temperatures below 39°C. Above this threshold temperature there was an approximate linear increase in evaporative heat
loss, by panting, with increase in POAH temperature. In the 15°C

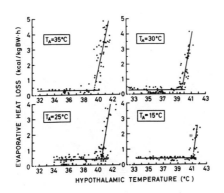

Fig. 2.   Relationship between hypothalamic temperatures
          and respiratory evaporative heat loss in a
          resting dog at four ambient temperatures (from
          Hellström and Hammel, 1967).

environment, lower right, there was a similar response to POAH
temperature but at a higher threshold of 41°C and, as might be
expected, a slightly steeper slope, an observation to be mentioned
later. These data, and many other evidences, clearly illustrate that
the POAH temperature has a way of affecting thermoregulatory
responses.

Effect of extrahypothalamic temperature transducers

A third very important property of the nervous system that I
would like to illustrate, before suggesting a neural network for
temperature regulation, is shown in Figure 3. For a prolonged
displacement of the POAH temperature from its normal level, there
is an initial, marked change in the amplitude of a thermoregulatory

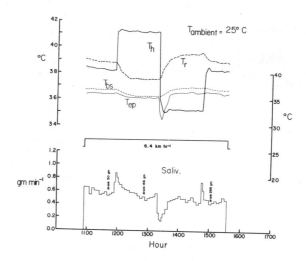

Fig. 3.   Salivation from parotid gland of dog while
running at 4 m.p.h. (6.4 km hr^{-1}) at 25°C.
Hypothalamic, T_h, and rectal, T_r, temperatures
are scaled on left. Back skin, T_{bs}, and ear
pinna, T_{ep}, temperatures are scaled on right.

response and there is thereby an initial imbalance between the rates
of heat production and heat loss. As one rate exceeds the other,
the core temperature changes accordingly and it changes at an
initial rate determined by the difference between the initial rates
of heat production and loss. In this record the thermoregulatory
response is salivation from one parotid duct in a dog running at
4 m.p.h. at 25°C. Heating the POAH tissue by 3°C caused a marked
initial increase in the rate of salivation while cooling 3°C caused
a marked decrease. As a consequence, the rectal temperature started
to decrease at a high rate or started to increase at a high rate
with heating or cooling of the POAH tissue, respectively. The
important third property of the nervous system is that the initial
responses are not sustained by steady displacement of the POAH
temperature. With changing core temperature, some property of
the nervous system involved in temperature regulation is also
changing. As this record illustrates, the core temperature changes
an amount so as to restore the system to a level of activity nearly
equivalent to the control level of activity. There must be other
temperature dependent neurons in the train of neurons which
regulated internal temperature.

Figure 4 will illustrate how these other neurons, somewhere in
the core, interact with the neurons in the POAH nuclei. The effect
of these other neurons is to change the threshold for activation
of a thermoregulatory response by the POAH tissue. The lower the
temperature of these other neurons, i.e. core temperature, upper
curve, the higher the POAH threshold temperature; or, conversely,
the higher the core temperature the lower the threshold, as shown
in the lower curve. These data, like those shown in Fig. 2, were

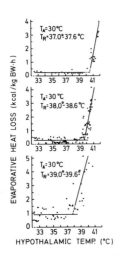

Fig. 4. Respiratory evaporative heat
loss versus hypothalamic
temperature in a resting dog
at 30°C ambient temperature
and at three levels of rectal
temperature (from Hellström
and Hammel, 1967).

the initial evaporative heat loss responses to displacement of
POAH temperature. Also, for each determination in Figure 4, the
rectal temperature was always at the level indicated, between 37.0
and 37.6°C for the upper curve, 38.0 and 38.6°C for the middle
curve and between 39.0 and 39.6°C for the lower curve.

A NEURONAL MODEL FOR TEMPERATURE REGULATION

A highly simplified version of a neural network which purports
to account for these first three properties of a mammalian temperature regulator is illustrated in Figure 5. In this figure one
neuron is depicted as performing the function of a population of
similar neurons, a singularly hazardous simplification. On the
left are depicted bipolar neurons in the dorsal root ganglia of
each spinal segment. These may also represent homologous neurons
in the trigeminal and facial ganglia. The afferent axons are
depicted as beginning in the cutaneous tissue as anatomically
undifferentiated free endings. The implication is that these
endings have the special property to transduce some feature of
the temperature profile near the surface into a generator potential
which is then propagated along the axon as action potentials. The
steady state temperature at the surface of the skin is shown to be
transduced to frequency of propagated action potentials by the "warm"

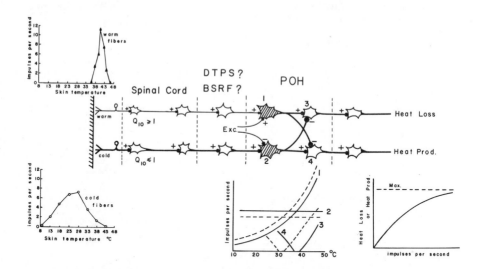

5.   A neuronal model for the regulation of body. temperature
in mammals. Heat Loss and Heat Prod. represent target
organs for affecting the rates of heat loss from the
body and heat production in such thermogenic tissue as
certain skeletal muscles, brown adipose tissue, etc.
Cell populations within ganglia and nuclei are represented by single neurons with only approximate
locations indicated or suggested. Four basic types are
indicated in the POH (preoptic-hypothalamic) tissue.
Neurons like 1 and 2 may or may not show spontaneous
activity in the absence of facilitation; they differ
markedly in the degree of temperature dependence, and
they can both facilitate and inhibit. Neurons like 3
and 4 have no spontaneous activity or temperature dependence and they receive more inhibitory than facilitory synapses for such costly responses as panting,
sweating or shivering. These neurons in the POH may
receive facilitory and inhibitory influence from neurons
in the diffuse thalamic projection system (DTPS) or the
brain stem reticular formation (BSRF), or both, and
these DTPS and BSRF neurons may be the links between
the peripheral temperature receptors and the POH as
well as between proprioceptors and the POH. Neurons like
1 are most concentrated within the preoptic-anterior
hypothalamic tissue whereas neurons like 2 may be widely
distributed throughout the hypothalamus and possibly
in the midbrain as well.

receptor as shown in upper left graph. Likewise, the "cold"
receptor transduces steady state surface temperature as shown
in the lower left graph. Since the warm receptors appear not to
function unless the skin temperature is extremly high (higher
than internal temperature) their role may be weakly indicated for

some animals. The afferent axons of the cutaneous temperature
neurons enter the spinal column and synapse with neurons in the
dorsal horn. In primates, these dorsal horn neurons are in the
proprius nucleus of each segment and their efferent axons cross
the commissure to enter the lateral spinothalamic tract on the
contralateral side. This tract then proceeds up the cord joining
the medial lemniscus in the medulla and then proceeds to the
thalamus, giving off collateral efferents to the brain stem reticular
formation (BSRF) in the medulla and mid-brain. In dogs and cats,
the efferent axons of the cutaneous temperature neurons enter the
cord and travel up the cord to synapse with dorsal horn cells in
the upper thoracic and lower cervical segments of the spinal cord.
Efferent axons from these dorsal horn neurons travel up the cord
in the spinocervical tract to terminate in the lateral cervical
nucleus on the ipsilateral side. Axons from this nucleus cross the
cord in the anterior commissure of the first and second cervical
segments to proceed forward and enter the medial lemniscus and end
in the thalamus, with collaterals to the reticular formation. The
spinocervical tract in cats and dogs with the added cervical
nucleus between the dorsal horn cells and the thalamus, probably
serve the same function as the spinothalamic tract in primates
(Breazile, 1970).

It is not clear how the neural traffic, conveying cutaneous
temperature information, passes to the preoptic and hypothalamic
nuclei even though it seems certain that connections do exist.
The principle sensory pathways for perceiving temperature effects
at the skin enter the ventrobasal complex of the thalamus and from
there radiate to the cortex to establish a point-to-point relationship between each receptive field in the skin and a specific somatic
sensory area in the cerebral cortex. It seems doubtful that this
thalamic-cortical specific projection system is utilized as a link
between the cutaneous receptors and the POAH nuclei for temperature
regulation. There exists another projection system in the thalamus,
the Diffuse Thalamic Projection System (DTPS), which projects from
the medial thalamic nuclei to the cortex, to the forebrain, preoptic and hypothalamic nuclei as well as to the mid-brain reticular
formation. Somesthetic stimuli proceeding to the thalamus also
affect this system. Perhaps the DTPS, either by way of the cerebral
cortex or directly, by way of the basal forebrain, is a link between
the cutaneous temperature transducers and the POAH tissue. Another
possible, perhaps probable, link is the brain stem reticular formation which extends from the medulla through the mid-brain to the
thalamus. It receives collaterals from all somesthetic receptors
and also from propricceptors in joints and muscle spindles. The
brain stem reticular formation (BSRF) is activated by these peripheral
receptors and it in turn activates the cerebral cortex to maintain
the state of wakefulness. The BSRF also has connections with the
hypothalamus and its activity almost certainly affects temperature
regulation in some way as will be suggested later.

At the present time I shall indicate, as shown in Figure 5,
that cutaneous sensory efferents join the POAH nuclei. They may
also synapse with other neurons in the regulatory network. In
the POAH nuclei there must exist a high concentration of neurons
that increase their activity with increasing temperature, or, to
say the same thing, they decrease their activity with decreasing
temperature. These are depicted by neuron 1 as having a Q_{10} of 2
(or more), that is, this neuron population exhibits an exponential
dependence upon temperature with a moderate activation energy.
Their distribution may not be exclusively within the preoptic nuclei

of the basal forebrain. Neurons in this population must facilitate
neurons of population 3 which activate heat loss mechanisms and at
the same time, inhibit neurons of population 4 which activate heat
producing organs.

In order that neurons of population 3 and 4 become active with
an appropriate dependence upon temperature, it is mandatory that
another population, depicted by neuron 2, be active and exhibit
a weakly positive, or zero or possibly negative temperature co-
efficient. Neurons like 2 must, at the same time, facilitate
neurons like 4 and inhibit neurons like 3. Only thus are neurons
like 3 active in an appropriate fashion to increase only those
thermoregulatory responses which increase the rate of heat loss as
the POAH tissue increases above a threshold temperature. Only thus
are neurons like 4 activated so as to increase the rate of heat
production as the POAH tissue decreases below a threshold temperature.

To insure that costly autonomic responses such as evaporative
heat loss and heat production have threshold temperatures that do
not overlap and are separated by an interval of 1 to 2°C, neurons
like 3 which activate the complex panting response (including salivation) or the sweating response, and neurons like 4 which activate
shivering and non-shivering thermogenesis are shown here to receive
slightly more inhibitory than facilitory fibers from neurons 1 and
2.

This combination of four neuron types has the basic properties to
be the controlling system of a closed loop, negative feedback,
temperature regulator with a set point. With the linkage between
the cutaneous temperature receptors and the hypothalamus, there
also exists the possibility to adjust the set by a command from
the periphery. The threshold for any thermoregulatory response
is subject to change by the external thermal environment acting
through the cutaneous receptors upon the temperature regulator.

As indicated in Figure 5, an increase in ambient temperature,
which will decrease the phasic and steady state activity in the
cutaneous "cold" receptors and which might increase the phasic
activity in the "warm" receptors, can react with neurons like 1
and 2 so as to decrease the threshold for all thermoregulatory
responses with only slight reduction in slope of response curves.
This effect occurs as one of the basic properties of the regulator.

The third in the list of basic properties to be explained was
the transient effect of displacing the POAH temperature. From this
result, we can infer that the activity of neurons in at least one of
the three nuclei between the dorsal root ganglia for the "cold
fibers" and the POAH nuclei is affected by its own temperature so
as to inhibit the effect induced by changing the POAH temperature.
This requires that the spinal cord neurons in the "cold fiber"
pathway exhibit a $Q_{10} < 1$. Cooling the cervical cord in the dog
(Simon et al., 1963 and 1965) and in the guinea pig (Brück and
Wünnenberg, 1965) facilitates shivering while heating the cord
facilitates panting (Jessen, 1967). These results may indicate
that cooling the spinal cord enhances the impulse traffic between
the cold receptors and the POAH nuclei so as to raise the threshold
for thermoregulatory responses whereas heating the cord diminishes
traffic over the same pathway so as to lower the thresholds. Thus,
prolonged heating of the POAH tissue which initially increases the
rate of heat loss also cools the spinal cord which could in turn

raise the threshold so that the response to heating is transient.
As the core temperature decreases-rapidly at first and then
gradually-to a lower, steady level so also does threshold for
heat loss response increase to a higher, steady level with the
result that the initial high heat loss activated by POAH heating
diminishes to about the same rate that prevailed before heating
commenced. The importance of this basic property of the neural
regulator of body temperature cannot be overemphasized. Its
importance has long been recognized in the effects of prolonged
POAH cooling and heating on the physiological rates of heat production and loss (Hammel et al., 1960 and 1963; Hellstrom and
Hammel, 1967; Andersson et al., 1963; Andersen et al., 1962). More
recently, prolonged heating or cooling of the POAH tissue causing
a decreased or increased core temperature, respectively, in turn
diminishes behavioral thermoregulatory responses initially activated
by POAH temperature displacement in the babboon (Gale et al., 1970)
and squirrel monkey (Adair et al., 1970). There can be little doubt
that extrahypothalamic temperatures in the core are effective in
some way, a fact that has been repeatedly emphasized by Thauer
(1970) and as early as 1935.

Let us examine how the next two properties of the nervous
system are to be incorporated into a neural network for regulation
of body temperature. First, how does exercise and then how does
hibernation affect thermal regulatory responses and can we infer
from these results how the neural network functions?

Effects of exercise induced inputs

Figure 6 shows how running at 0, 2, 4 and 6 m.p.h. affects the
relationship between salivation and the POAH temperature. The
environmental temperature was 25°C in all experiments. Two
striking effects are noticed. The threshold for thermoregulatory
salivation was decreased by each increase in the level of exercise.
In fact, while running at 4 and 6 m.p.h. at 25°C, cooling the POAH
tissue to 31°C was not sufficient to suppress salivation. The
other effect to notice is that the slopes of the response curves
were diminishing with each increase in the level of exercise. It
would appear that thermoregulatory salivation is a saturable
response, that is, as it approaches its maximum response, increasing
the neurological stimulus yields a diminishing response.

The same two effects can be seen in the four experiments shown
in Figure 7. where the running rate was 4 m.p.h. in all cases while
the ambient temperature was maintained at four levels 5, 15, 25 and
35°C. As the external thermal stress increased, the threshold for
salivation decreased and the slopes of the response curve were
flattening, again suggesting saturation in the response of the
parotid gland.

How are these two effects to be incorporated into a model of
a neural network for temperature regulation? There seems to be
no difficulty in supposing that some, if not all, thermoregulatory
responses have a maximum capability and that the maximum is approached by a non-linear relationship between stimulus and response,
as suggested in the lower right corner of Figure 5. There is also
the possibility that the response to facilitory and inhibitory
inputs of neurons like 1, 2, 3 and 4 are all non-linear. These
possibilities are indicated only slightly in Figure 5. The important probable fact is that thermoregulatory responses are less
than linear responses at the highest levels of POAH stimulation.

130 H. T. HAMMEL

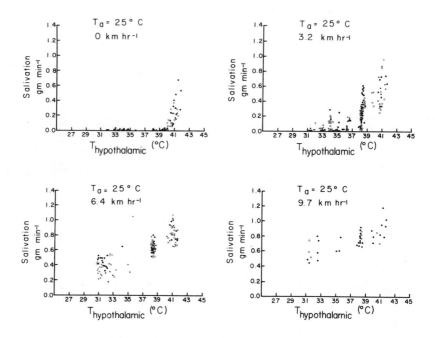

Fig. 6.   Rates of salivation from one parotid gland
          versus the hypothalamic temperature, while
          running at 0, 2, 4 and 6 m.p.h. (miles per
          hour) at 25°C (from Hammel and Sharp, 1971).

     The effect of exercise upon threshold can be accounted for by
suggesting that the neural traffic from proprioceptors which is
relayed up the spinal cord and brain stem to the thalamus and with
collaterals to the reticular formation is also affecting neurons
like 1 or 2 in the POAH nuclei. This proprioceptor input could
either facilitate neurons like 1 or inhibit units like 2, or both.
Either way, the threshold for all thermoregulatory responses would
be lowered by exercise. Also, since the activity of neurons like
1 is presumed to depend exponentially upon temperature, inhibition
of neurons like 2 could explain the observation that changing the
ambient temperature yields a larger effect on threshold in the
running dog than in the resting dog. That is to say, if changing
the environmental temperature from 35° to 15°C affects the level
of activity of neurons like 2, that effect would decrease the
thresholds less in the resting animal than in the running animal
because the activity of neurons like 2 is already reduced by the
exercise. At the very least, the experimental evidence does not
contradict this interpretation. Also, as pointed out earlier, this
interpretation of the effects of peripheral input from cutaneous
receptors upon neurons like 2 might explain the observation that
the slope of evaporative heat loss vs. hypothalamic temperature
was slightly steeper in the cold environment, Figure 2, and slightly
steeper for the low core temperature, Figure 4. The explanation
may be that as neurons like 2 are facilitated by a hypothermic skin

and core, the net interaction of neurons like 1 (with their

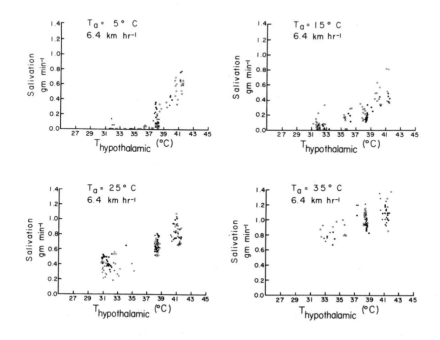

Fig. 7. Rates of salivation from one parotid gland
versus the hypothalamic temperature, while the
dog ran at 4 m.p.h. at four ambient temperatures
5, 15, 25 and 35°C (from Hammel and Sharp, 1971).

exponential temperature dependence) and 2 upon neurons like 3 as a
function of POAH temperature would be steeper than when neurons like
2 are not facilitated.

Effects of torpor related inputs

     Next, how does hibernation affect the relationship between POAH
temperatures and thermoregulatory responses? Suppose the brain
stem reticular formation, receiving collaterals from all somesthetic
and proprioceptor sensory inputs, is one of the links between peripheral information and the POAH nuclei. Activity in the BSRF could
then be profoundly influential in sustaining and adjusting levels
of activity in neurons like 1 and 2. Further, suppose that in
hibernation the BSRF is depressed deeply. All temperature regulation would likewise be depressed. Changing the temperature of the
POAH tissue would have no measurable effect on oxygen consumption
or on any other thermoregulatory response. As shown in Figure 8,
displacing the POAH temperature from 3 to 13°C elicited no response
in the oxygen consumption of a hibernating ground squirrel implanted
with thermodes. From this, we have concluded that the central

nervous regulator of body temperature is inactivated during hibernation. However, the regulator is capable of reactivation with appropriate stimuli. A slowly decreasing air temperature will arouse
a golden mantled ground squirrel from deep hibernation when a

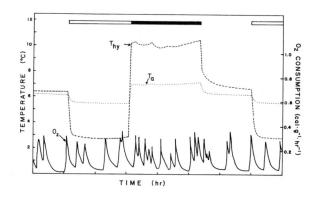

Fig. 8.   Absence of effect of heating and cooling preoptic
           tissue upon the periodic oxygen extraction curve
           for a hibernating ground squirrel Citellus
           lateralis.

certain lower alarm temperature is reached, and an arousal so induced
can be reversed if the air temperature is raised soon after arousal
begins. This experiment is shown in Figure 9. An arousal was induced
by slowly lowering air temperature to 2°C. The arousal was reversed
by heating the preoptic tissue or some adjacent part of the brain
stem. The preoptic heating was continued as air temperature was
lowered below the level which induced the arousal. When the hypothalamic temperature was reduced by half a degree, another arousal
began which was again suppressed by increasing the level of heating
of hypothalamic temperature.

    Animals hibernating at air temperatures several degrees above
their lower alarm temperature can be induced to arouse by briefly
cooling hypothalamic temperature, as shown in Fig. 10. In this
experiment the ambient temperature was 6.5°C. Cooling the preoptic
tissue to 2°C and heating to 10°C elicited no response in oxygen
consumption, other than to change the pattern of respiration and
oxygen extraction when cooling the brain stem below about 3°C, a
change that had been previously noted when induced by ambient cooling
(Hammel et al., 1968). On the other hand, arousals were twice
induced by cooling the preoptic tissue to 0.8°C and twice reversed
by heating to 12.5°C.

    The record in Figure 11 began during an arousal from deep
hibernation and the arousal was in an advanced stage. Note that the
scale for oxygen consumption on the right is 20 times the scale
used in Figure 10 and 40 times that in Fig. 8. When hypothalamic
and internal body temperatures reached 30°C, the preoptic tissue was
heated to 39°C. Immediately, the high rate of heat production

associated with arousal plummeted to basal level and an internal body
temperature dropped from 30 to 22.5°C during 30 minutes of preoptic
heating. The preoptic tissue was then cooled to 28°C and held
between 28 and 30° for 45 minutes. •During this time the body
temperature rose from 22.5 to 41.6°C at an average rate of 0.48°C/min.
Body temperature normally plateaus at 36 to 37°C following an arousal
from torpor and it returned to that level when preoptic cooling was

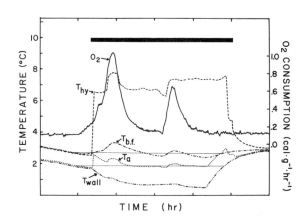

TIME (hr)

Fig. 9.   Effects of an arousal upon O_2 consumption induced
by slowly cooling the ambient temperature of a
hibernating ground squirrel. Note that the
arousal may be reversed by warming some part of
the brain stem (from Heller and Hammel, 1972).

stopped. The rather steep rise of body temperature to such a high
level and still climbing and the sustained maximum rate of heat
production during preoptic cooling indicate that there is little
negative feedback from temperature dependent neurons outside the
preoptic region in this animal.

    Such high sensitivity of the preoptic anterior hypothalamic
tissue to temperature displacements and such low negative feedback
has also been found in the non-hibernating season in our laboratory.
For example, in a typical experiment at an air temperature of 13°C
and using a ground squirrel in breeding condition, the preoptic area
was cooled 1.0°C below the threshold level of 38.5°C and clamped
there for 22 minutes. O_2 consumption rapidly increased threefold
and body temperature rose at an initial rate of 0.54°C/min. At the
end of 22 minutes of cooling, the body temperature was 42.5°C and
still rising. The proportionality constant for activation of
shivering in the squirrel was approximately 15 times that for the
dog and the open loop gain for the squirrel was approximately 70,
whereas for the dog it is 15 (Hammel, 1968).

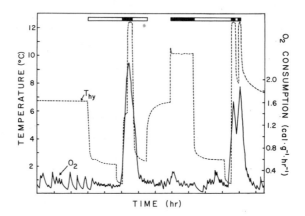

Fig. 10.   Effects of an arousal upon O₂ consumption
           induced by cooling some part of the brain
           stem to below an alarm temperature (from Heller
           and Hammel, 1972).

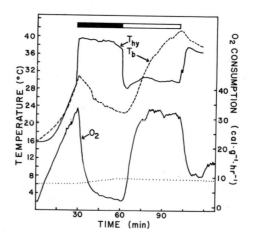

Fig. 11.   Effects of heating and cooling the hypothalamic
           tissue during an advanced stage of arousal when
           the ground squirrel was near normothermic core
           temperatures (from Heller and Hammel, 1972).

A plausible interpretation of the radical and reversible alteration of the regulator of body temperature for entry into and arousal
from hibernation may be suggested by reference again to the model
in Figure 5. If those neurons indicated here to reside in the brain
stem reticular formation or in the diffuse thalamic projection

system   sustain the activity of neurons like 2 so that normal
temperature regulation is achieved in the non-hibernating squirrel,
then loss of this activity could lead to inactivation of temperature
regulation during hibernation.   It would appear to us that when
the temperature of these neurons in the BSRF is above alarm
temperature, the regulator is inactive during hibernation.
When they drop below alarm temperature, which is less than 1°C,
then regulation is reactivated.   It would also appear that the
hibernating animal receives very little input from the peripheral
cold receptors during hibernation; and in the euthermic state, the
neurons in the spinal cord subserving peripheral thermal receptors
do not respond to temperature.   We might suppose that adaptations
of the temperature regulator of hibernators include 1) a stepwise
reduction in responsiveness of those neurons in the BSRF which
receive facilitation from the collateral fibers of the peripheral
temperature sensing pathways - a partial reduction in responsiveness
characterizing the sleep state; total inactivation characterizing
hibernation; and reactivation by a near 0°C alarm temperature
characterizing arousal, 2) neurons in the spinal pathway between the
cutaneous cold receptors and the POH which have a $Q_{10} = 1$, and
3) cutaneous cold receptors which are inactive at temperatures near
0°C.   Finally, we might also suppose that the hypothermia induced by
anesthetics results from the inactivation of these same neurons in
the BSRF or DTPS which in the wakeful state facilitate the
controlling neurons in the POH.   Without such facilitation, the
POH neurons may not function.

SUMMARY AND LIMITATIONS

     The salient features that appear to me to be relevant for
guessing at a neural network for regulation of internal body
temperature in endotherms are included in Figure 5.   A similar
network may also apply to ectothermic vertebrates from fish to
reptiles where the only changes would be to replace "Heat Loss"
with "Exit from Hot Environment" and replace "Heat Production"
with "Exit from Cold Environment" when these two behavioral
responses are possible.   In both cases the effects are to balance
the rates of heat loss and heat production at some body temperature
that is near to the optimal temperature for some body functions.
Even more sophisticated responses may be activated in reptiles such
as basking, orientation, shade seeking, voiding cloacal fluid,
burrowing, etc.   In fact, these and other behavioral responses
may often be employed and preferred by endotherms as well.   There
is increasing evidence that these responses in all vertebrates
are controlled by a controlling system with the same characteristics
as are outlined in Figure 5, and with only minor alterations.

     The concept of the set point temperature as applicable to
temperature regulation in vertebrates is defensible.   It may be
defended as an analogy that yet awaits the status of proven reality.
At the same time I cannot view the set point temperature as an
invariant property of the regulator.   Indeed, thresholds for all
thermoregulatory responses are readily shifted by many inputs.

     Some limiting comments may render this position more tenable.
A simplified model, such as this, does not exclude other features,
other neurons or other pathways.   Rather it purports to include
only the salient features, neurons and pathways.   A model like this
has some small predictive value and some interpretive value.   One

could not predict the activity of all the single units to be found
between the basal forebrain and the mid-brain in a conscious animal
of a given species subjected to a variety of external or internal
thermal stresses. Especially misleading would be a prediction of
unit activities in a brain stem inactivated by anesthesia. Furthermore, since the relationship between the neurological signal and
the thermal regulatory response of a gland or muscle is poorly
known, it becomes impossible to predict a precise relationship
between the hypothalamic temperature and a thermoregulatory response.
All that we have attempted to do is measure the responses activated
by controlling the hypothalamic temperature under several sets of
conditions and then interpret these responses in terms of a plausible
model involving neurons. Therefore, I am not surprised that this
simple model fails to predict the electrical activity of all the
neurons in the brain stem or fails to predict some of the results
obtained by infusing transmitter substances into the rostral brain
stem.

REFERENCES

Adair, E. R., J. V. Casby and J. A. J. Stolwijk, (1970), Behavioral
    temperature regulation in the squirrel monkey: changes induced
    by shifts in hypothalamic temperature, J. Comp. Physiol. Psychol.
    72, 17.
Andersen, H. T., B. Andersson and C. Gale, (1962), Central control
    of cold defense mechanism and the release of "endopyrogen" in
    the goat, Acta Physiol. Scand. 54, 159.
Andersson, B., L. Ekman, C. C. Gale and J. W. Sundsten, (1963),
    Control of thyrotrophic hormone (TSH) secretion by the "heat loss
    center", Acta Physiol. Scand. 59, 12.
Breazile, J. E., (1970), Chapt. 42, Spinal cord and brain stem
    function, In: Duke's Physiology of Domestic Animals, editor:
    M. J. Swenson, (Cornell University Press, Ithaca), p. 918.
Brück, K. and W. Wünnenberg, (1965), Beziehung zwischen thermogenese
    im "braunen" fettgewebe, temperatur im cervicolen anteil des
    vertebrakanals und kaltezittern, Arch Ges. Physiol. 290, 167.
Gale, C. C., M. Mathews and J. Young, (1970), Behavioral thermo-
    regulatory responses to hypothalamic cooling and warming in
    babboons, Physiol. Behav. 5, 1.
Hammel, H. T., (1968), Regulation of internal body temperature,
    Ann. Rev. Physiol. 30, 641.
Hammel, H. T., T. J. Dawson, R. M. Abrams and H. T. Andersen, (1968),
    Total calorimetric measurements on Citellus lateralis in hiberna-
    tion, Physiol. Zool. 41, 341.
Hammel, H. T., J. D. Hardy and M. M. Fusco, (1960), Thermoregulatory
    responses to hypothalamic cooling in unanesthetized dogs, Am. J.
    Physiol. 198, 481.
Hammel, H. T., D. C. Jackson, J. A. J. Stolwijk, J. D. Hardy and
    S. B. Strømme, (1963), Temperature regulation by hypothalamic
    proportional control with adjustable set temperature, J. Appl.
    Physiol. 18, 1146.
Hammel, H. T. and F. Sharp, (1971), Thermoregulatory salivation in
    the running dog in response to preoptic heating and cooling, J.
    Physiol. (Paris) 63, 260.
Heller, C. and H. T. Hammel, (1972), CNS control of body temperature
    during hibernation, J. Comp. Biochem. Physiol. (in press).

Hellstrøm, B. and H. T. Hammel, (1967), Some characteristics of
    temperature regulation in the unanesthetized dog, Am. J. Physiol.
    213, 547.
Iggo, A., (1970), Chapt. 43, Somesthetic sensory mechanisms, In:
    Duke's Physiology of Domestic Animals, editor: M. J. Swenson,
    (Cornell University Press, Ithaca) p. 949.
Jessen, C., (1967), Auslosung von hecheln durch isolierte warmung
    des ruckenmarks am wachen hund, Plugers Archiv 297, 53.
Simon, E., W. Rautenberg and C. Jessen, (1965), Initiation of
    shivering in unanesthetized dogs by local cooling within the
    vertebral canal, Experienta 21, 477.
Simon, E., W. Rautenberg, R. Thauer and M. Iriki, (1963), Auslosung
    thermoregulatorischer reaktionen durch lokale kuhlung im
    vertebralkanal, Naturwissenschaften 50, 337.
Thauer, R., (1970), Temperature reception in the spinal cord, In:
    Physiological and Behavioral Temperature Regulation, editors:
    J. D. Hardy, A. P. Gagge and J. A. J. Stolwijk, (Charles C.
    Thomas, Springfield, Ill.).

    Preparation of this manuscript was supported in part by U.S.
Public Health Service Grant 1 R01 GM 17222-01 and Contract AF
F33615-69-C-1024 with Wright-Patterson AFB.

JAN W. SNELLEN
Memorial University of Newfoundland
St. John's, Newfoundland, Canada

Most models of temperature regulation during exercise are based on central and peripheral thermal inputs to the control centre. Robinson (1949) was the first one to show the interaction between sweat rate as the controlled output on one hand and rectal (as representative of body core) temperature and mean skin temperature on the other hand (fig. 1).

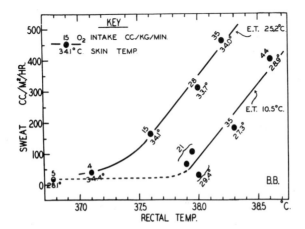

Fig. 1.   Sweat rate as a function of rectal and mean skin temperature at various levels of exercise and ambient conditions (Effective Temperature). By courtesy of Dr. Robinson and W. B. Saunders Publishing Co.

This basic model has since then been modified in various ways.

a) Rectal temperature is a rather slowly responding representative of core temperature and is, for some obscure reason, higher than any other core temperature. As a result, esophageal and tympanic membrane temperatures have been introduced as alternative methods of measuring body core temperature.

b) The straight lines in fig. 1 have been replaced by curved lines (Benzinger, 1961).

c) The straight lines running parallel have been replaced by a set of lines originating from one point, either as a set of straight lines fanning out (Stolwijk and Hardy, 1966; Nadel, et al., 1971) or as a set of S-shaped curves (Wyndham, 1965).

d) The relationship between sweat rate or metabolism, as another controlled output, and the two temperatures has been expressed in mathematical form (Hardy and Stolwijk, 1966; Stolwijk and Hardy, 1966; Nadel, et al., 1971).

Some authors weigh core and average skin temperatures and add them together,

others multiply temperatures and obtain equations where sweat rate is a function
of a temperature squared. Polynomial equations with temperature squared have
also been proposed (Kitzing, et al., 1971).

The underlying philosophy of all these models is that the thermoregulatory
mechanism has two input signals: average skin temperature and body core temperature, presumably sensed by the hypothalamus. The temperature of this latter
area, being inaccessible in human thermoregulation studies, is represented by
tympanic membrane (or worse, outer ear hole) temperature.

The rationale of this philosophy is the idea that skin temperature represents the external heat or cold load to the controlled system and core temperature
the internal heat load. It has been demonstrated clearly by M. Nielsen (1938)
that skin temperature is only a function of air temperature, independent of work
load, and core temperature (rectal in Nielsen's studies) a linear function of
metabolic rate, independent of ambient temperature. Although Nielsen's name has
been misspelled in the literature as Neilsen, Nielson or Neilson, his work has
been confirmed to a great extent by various later investigators, but with the
following amplifications:

a) Skin temperature is a function of air temperature (B. Nielsen, 1969; Snellen,
    1966) up to a value of about $35^{\circ}C$, when it tends to level off (Mitchell,
    et al., 1968). It seems that acclimatisation to heat improves this levelling
    off.

b) Local application of radiant heat operates as a thermosensitive amplifier.
    When the average skin temperature is $35^{\circ}C$, and thus independent of work load
    and ambient temperature, a given rise in local skin temperature due to radi-
    ation multiplies the already existing local sweat rates (varied by exercise
    and/or high ambient temperatures) by a constant factor. This rise in sweat
    rate is in no way related to the amount of heat irradiated, but only to the
    local skin temperature (Mitchell and van Rensburg, 1971).

c) Core temperature is affected, to a small degree, at low work loads, by ambient
    conditions (Kitzing, et al., 1968).

d) Interindividual differences in the linear relationships between core temper-
    ature and metabolic rate (V_{O_2}) can be eliminated to a great extent when

    oxygen consumption is expressed as a percentage of the maximum oxygen uptake
    (I. Åstrand, 1960; Saltin and Hermansen, 1966).

Thus the thermoregulatory model with two temperature inputs seemed to be
based on solid grounds: one temperature for external heat loads and one temperature for internal heat loads.

More serious amendments to this two-temperature model emerged when it became
clear that core temperature was not a linear function of the heat produced in the
body, but only of oxygen consumption. This observation arose from experiments in
which the external work during exercise was varied. Snellen (1960) showed
(fig. 2) that sweat rate could be changed without changing mean skin temperature,
or rectal temperature or oxygen consumption, by only varying the external work
done on a motor driven treadmill. The total heat production during walking uphill
was about 1 kcal/min less than during level walking, and this difference was
reflected precisely by the difference in the evaporative heat loss. A systematic
study of the effect of varying metabolic rate and positive and negative work load
was carried out by B. Nielsen (1966). Her results can be summarised as follows:

a) She confirmed previous observations (Abbott, et al., 1952; Asmussen, 1952)
    that during negative work (by cycling and running downhill), the oxygen con-
    sumption is much less than during positive work, with numerically the same

external power, but with reversed sign.

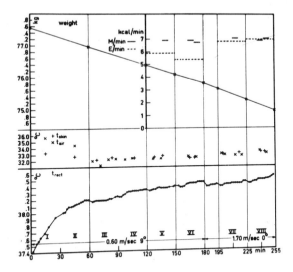

Fig. 2.   Three hours walking uphill and one hour on the level with the same
           metabolic rate. Mean skin temperature and air temperature kept
           equal. Evaporative heat loss (E/min) changes when external work
           is changed. By courtesy of J. appl. Physiol.

b)   Core temperature followed closely the metabolic rate $\dot{M}$ (in B. Nielsen's
     terminology, metabolic heat liberation; in Snellen's terminology, gross meta-
     bolic heat production) but had no relationship to the total heat generated in
     the body ($\dot{M} + \dot{W}$, where $\dot{W}$ is the work load, with a negative sign during
     positive work and a positive sign during negative work--in B. Nielsen's
     terminology, total heat production; in Snellen's terminology, net heat pro-
     duction).

c)   Sweat rate followed closely $\dot{M} + \dot{W}$, but was not related to esophageal temper-
     ature.

     These observations cast serious doubt on the validity of the concept that
core temperature is the proper representative of the internal heat load. If mean
skin temperature is considered to reflect external heat load, it seems better to
replace T_s by the difference between air temperature and mean skin temperature
$(T_a - T_s)$, which is, within limits and at one wind velocity, proportional to the
sum of radiative and convective heat exchange $\dot{R} + \dot{C}$. Also, if core temperature
is not representative of internal heat load, it may be replaceable by $\dot{M} + \dot{W}$.

     The result of this manoeuvre (fig. 3), however, is a triviality, because
after conversion to the same units it is nothing more than the statement that
during caloric equilibrium the heat production equals heat loss, or $\dot{M} + \dot{W} + \dot{R} + \dot{C}$
$= -\dot{E}$ (positive sign:heat gain by the body, $-\dot{E}$ = evaporative heat loss), which is
the first law of thermodynamics, written in physiological terms.

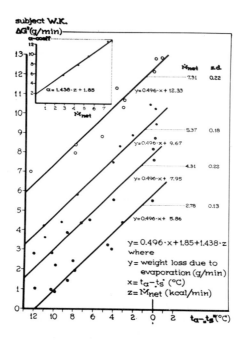

Fig. 3. Sweat rate as a function of external heat load, represented by
$T_a - T_s$, and internal heat load, represented by $\dot{M} + \dot{W}$ (net heat
production). By courtesy of Acta physiol. pharm. neerl.

More doubt about the validity of a two-temperature model has been presented
by Snellen (1966). He used direct calorimetry, and summed all heat fluxes to
and from the body during a 60 minute recovery period after exercise until a final
steady state was reached (fig. 4). He thus obtained the negative heat storage
in the recovery period which was, as far as he could estimate, identical to the
heat accumulated during the period of work and heat load. This heat storage was
divided by the water equivalent of the body (0.83 times body weight) and yielded
the change in average body temperature during exposure to work and heat load.

In Snellen's experiments the subjects rested in a calorimeter, left the
calorimeter and exercised in a climatic room, and then returned to the calorimeter. His study included all combinations of three calorimeter temperatures,
four work loads and four ambient temperatures during exercise. His results can
be summarised as follows:

a) The change in average body temperature shows a close relationship with sweat
    rate during exercise (fig. 5). The sweat rate is, in the range investigated,
    directly proportional to the change in average body temperature. At a low
    calorimeter temperature (30°C) slight lowering of the average body temperature
    would result in zero sweating. At a high calorimeter temperature (40°C) a
    greater decrease in mean body temperature would be required to stop sweating.
    Sweat rates while sitting in the calorimeter, during caloric equilibrium,
    i.e., at zero change in average body temperature (indicated by crosses),
    coincide with the observed curves. This is evidence that there is no set
    point shift (see below) during exercise. The slope of the signal-response

curve is different for various subjects.

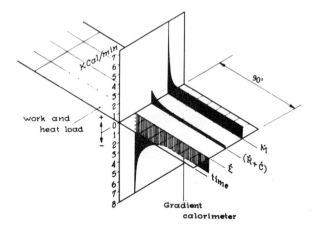

Fig. 4. Calorimetric determination of body heat storage. The sum of the
         three shaded areas represents the heat storage during the exercise
         period (prior to time 0). By courtesy of Medicine and Science in
         Sports.

b) The change in average body temperature cannot be represented validly by
    observed changes in mean skin and rectal temperature. Regression analysis
    revealed rectal temperature to contribute more than 100%.

c) Rectal temperature was firstly a function of metabolic rate $\dot{M}$ rather than
    $\dot{M} + \dot{W}$ and secondly a function of dehydration.

     Gagge and Saltin (1971) have recently demonstrated, by using heat exchange
equations rather than direct calorimetry to calculate average body temperature,
that a) is also valid in the acute transient stage from work to rest. Result b)
is a confirmation of the conclusion reached above that a two-temperature model is
doomed to failure. If average body temperature controls sweat rate, and if
average body temperature cannot be predicted from the two "classical" temperatures, then

1) the integrative centres must receive more thermal information from the body
    than we can obtain with probes at the core and the skin, and

2) there must be a large body mass with a temperature higher than rectal temper-
    ature.

     There is ample evidence that the latter is true. Working muscles have a
temperature higher than rectal temperature (Robinson, et al., 1965; Saltin and
Hermansen, 1966; B. Nielsen, 1969; Saltin, et al., 1970). During negative work,
when work is done on the stimulated muscle, the muscle temperature is even higher
than during positive work (B. Nielsen, 1969). No direct comparison has been made
between average body temperature obtained calorimetrically and mean skin temperature, working and non-working muscle temperatures and core temperature. "Core"
temperature may be the temperature of a relatively small fraction of the body
(Snellen, 1969a).

     The former conclusion, namely, that the integrative centres receive more
thermal information than can be obtained from thermometers, cannot be demon-

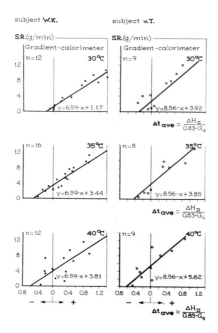

Fig. 5.   Sweat rate as a function of the change in average body temperature,
          as determined calorimetrically.   By courtesy of Acta physiol.
          pharm. neerl.

strated directly from experiments on humans.   There is indirect evidence, however, from experiments on humans: any heat flux towards the body is counter-
acted, after some delay, during which a heat storage can be built up, by an active
response which induces an equal heat flux from the body, irrespective of the place
of application of the heat load.   This holds true for radiant heat load to the
skin during rest (Gagge and Hardy, 1967), passive deep body heating by diathermia
(B. Nielsen and M. Nielsen, 1965), heat flux to the muscle during negative work
(B. Nielsen, 1966), heat production in the body due to various types of exercise,
and a heat load applied to the stomach (Snellen, et al., 1971).

The analogy of a bucket (Mitchell, this book) with variable water influx
(heat load) and an outlet (evaporative heat loss) which is controlled by a float
(water level ≡ heat storage) is obvious.   To conclude, however, that the content
of the bucket is regulated, or that there is heat regulation rather than temperature regulation, is hazardous.   Heat regulation requires not only information
about the temperature, but also information about the volume.   Yet there are
indications that the body behaves as if heat regulation is present.   This does
not imply constancy of body heat content.   Body heat content is a misleading
term, especially when the exercising body losses mass by dehydration (Snellen,
1969b).

It is a well-established fact that during dehydration the rectal temperature
at a particular metabolic rate is higher than during a state of normal hydration,
and lower during hyperhydration (Pitts, et al., 1944; Moroff and Bass, 1965).
The dehydration effect on rectal temperature was also found by Snellen (1966,
fig. 6, left).   Even average body temperature (heat storage divided by the water

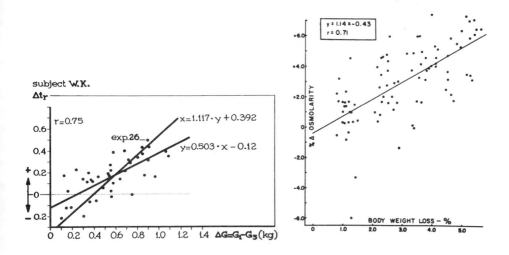

Fig. 6.   The effect of dehydration on rectal temperature (left) and on
          plasma osmolarity (right). Left graph by courtesy of Acta physiol.
          pharm. neerl. and right graph by courtesy of Dr. L. C. Senay and
          J. appl. Physiol.

equivalent of the body, $\Delta H/0.83$ G) must rise in dehydrating subjects, even during
caloric equilibrium, when rate of change in storage is nil, because body weight
G decreases.

On the other hand, plasma osmolarity rises during dehydration (Senay and
Christensen, 1965, fig. 6, right). Administration of large doses of hypertonic
saline affects both plasma osmolarity, circulating blood volume and esophageal
temperature (B. Nielsen, et al., 1971). If plasma osmolarity could be regarded
as a derivative of body volume, and if plasma osmolarity and average body temperature could interact as a constant relationship (constant quotient), heat
regulation could be achieved without constancy of body heat content. At the
moment this idea is highly speculative, but not more speculative than the proposition by Myers and Veale (1970) that the set point for thermoregulation depends
on the sodium-calcium ratio in the CSF, nor more fantastic than the proposition
that sweat rate is controlled by a temperature squared.

Aside from this speculative point, dehydration during work in the heat is of
paramount importance both from a practical (marathon running, long distance
cycling) and a theoretical point of view. Senay (1971) has found that, during
dehydration, during exercise and during heat exposure, body water, electrolytes
and proteins are redistributed between body compartments to an extent not previously recognized.

The designated title of this lecture is "set point during exercise." As we
have pointed out elsewhere (Mitchell, et al., 1970) set point is a concept
derived from control theory and should be thought of as the input value of a
control system at zero output. We argued that there is not such a thing as an
anatomical structure providing a reference signal to the integrating centres of
the thermoregulatory mechanism. We deduced from neuronal behaviour that it is
more likely that the "set point" is achieved by a balance between integrated cold
and integrated warm input (fig. 7).

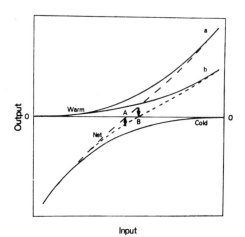

Fig. 7.   "Set-point" A is achieved by a balance between integrated warm
          input (upper solid line a) and integrated cold input (lower solid
          line). The resulting input-output relationship is the upper
          broken line. A change in sensitivity (lower solid line b) results
          in a "set-point shift" from A to B. By courtesy of Pflüg. Arch.

    No shift, even in apparent "set point," has to be assumed during exercise
(see above). A possible interaction between osmolarity, or Na:Ca ratio, and
temperature does not necessarily imply that osmolarity, or Na:Ca ratio, acts as
the set point for thermoregulation. It is equally possible that a change in
osmolarity, or Na:Ca ratio, changes the sensitivity of the neurones in the
thermoregulatory pathways.

    The elevation of core temperature during exercise is classically described
as a "set-point shift." If we reject this concept it is necessary to provide an
alternative explanation. The answer seems to lie in the regional redistribution
of blood flow during exercise. Cardiac output and muscle blood flow during
exercise are functions of oxygen consumption and are, to a large extent,
independent of ambient conditions. However, in hot environments skin blood flow
is increased drastically to facilitate heat transport to the body surface.
Therefore, either there is an increase in circulating blood volume or a redistribution of regional blood flow. Since it is known that in acute exposure to
work and heat blood volume actually decreases, redistribution must occur. This
has been demonstrated by Rowell, et al. (1970), who showed a decrease in
splanchnic-hepatic blood flow which was the mirror image of the rise in rectal
temperature. If splanchnic-hepatic heat production (Grayson, 1971) is not
affected by exercise, and if splanchnic-hepatic perfusion decreases, a rise in
temperature in this area must result. According to this reasoning "core"
temperature inversely reflects splanchnic-hepatic blood flow, which decreases
as a result of a demand for blood from other areas. It satisfies the observations that rectal temperature is also affected by ambient temperatures and that
dehydration, with a concomitant decrease in circulating volume, causes an extra
rise. It also meets the observation that running dogs do not seem to have a
rise in rectal temperature (Jackson and Hammel, 1963); in these animals, which do
not sweat, an increased skin blood flow is redundant.

    In conclusion, there is ample evidence that a thermoregulatory model, based
on one central and one peripheral input is an oversimplification. If average

body temperature is the proper input signal, there is no need to assume a set point shift during exercise. The regulating mechanism operates as if there is heat regulation instead of temperature regulation. This may be achieved if an interaction between blood osmolarity (as a derivative of body mass) and average body temperature can be assumed. Rectal temperature and skin temperature alike are passive results of a local heat balance.

REFERENCES

Abbott, B. C., B. Bigland and J. M. Ritchie (1952), The physiological cost of negative work, J. Physiol., Lond. 117, 380.

Asmussen, E. (1952), Positive and negative muscular work, Acta physiol. scand. 28, 364.

Åstrand, I. (1960), Aerobic work capacity in men and women with special reference to age, Acta physiol. scand. 49, Suppl. 169.

Benzinger, T. H. (1961), The diminution of thermoregulatory sweating during cold-reception at the skin, Proc. nat. Acad. Sci., Wash. 47, 1683.

Gagge, A. P. and J. D. Hardy (1967), Thermal radiation exchange of the human body by partitional calorimetry, J. appl. Physiol. 23, 248.

Gagge, A. P. and B. Saltin (1971), Prediction of human body temperature during maximal exercise from skin, rectal, esophageal and muscle temperatures, Proc. XXV I.U.P.S. Congr., München.

Grayson, J. and A. O. Durotoye (1971), Effect of environmental temperature on gastrointestinal heat production, Abstr. int. Symp. environ. Physiol.: Bioenergetics and temp. Regul., Dublin.

Hardy, J. D. and J. A. J. Stolwijk (1966), Partitional calorimetric studies of man during exposures to thermal transients, J. appl. Physiol. 21, 1799.

Jackson, D. C. and H. T. Hammel (1963), Reduced set point temperature in exercising dog, Tech. Doc. Rep. AMRL-TDR-63-93.

Kitzing, J., K. Behling, A. Bleichert, S. Ninow, M. Scarperi and S. Scarperi (1971), Correlations between input and output of the thermoregulatory control system during rest and exercise, Abstr. int. Symp. environ. Physiol.: Bioenergetics and temp. Regul., Dublin.

Kitzing, J., D. Kutta and A. Bleichert (1968), Temperaturregulation bei langdauernder schwerer Körperlicher Arbeit, Pflüg. Arch. 301, 241.

Mitchell, D. and A. J. van Rensburg (1971), Thermoregulatory significance of the effect of local skin temperature on local sweat rate, Proc. XXV I.U.P.S. Congr., München.

Mitchell, D., J. W. Snellen and A. R. Atkins (1970), Thermoregulation during fever: change in set-point or change in gain, Pflüg. Arch. 321, 293.

Mitchell, D., C. H. Wyndham, A. R. Atkins, A. J. Vermeulen, H. S. Hofmeyr, N. B. Strydom and T. Hodgson (1968), Direct measurements of the thermal responses of nude resting men in dry environments, Pflüg. Arch. 303, 324.

Moroff, S. V. and D. E. Bass (1965), Effects of overhydration on man's physiological responses to work in the heat, J. appl. Physiol. 20, 267.

Myers, R. D. and W. L. Veale (1970), Body temperature: possible ionic mechanism in the hypothalamus controlling the set point, Science 170/3953, 95.

Nadel, E. R., S. M. Horvath, C. A. Dawson and A. Tucker (1971), Sensitivity to central and peripheral thermal stimuli in man, J. appl. Physiol. 29, 603.

Nielsen, B. (1966), Regulation of body temperature and heat dissipation at different levels of energy- and heat production in man, Acta physiol. scand. 68, 215.

Nielsen, B. (1969), Thermoregulation in rest and exercise, Acta physiol. scand., Suppl. 323.

Nielsen, B. and M. Nielsen (1965), Influence of passive and active heating on the temperature regulation in man, Acta physiol. scand. 64, 323.

Nielsen, B., G. Hansen, S. O. Jørgensen and E. Nielsen (1971), Thermoregulation in exercising man during dehydration and hyperhydration with water and saline, Abstr. int. Symp. environ. Physiol.: Bioenergetics and temp. Regul., Dublin.

Nielsen, M. (1938), Die Regulation der Körpertemperatur bei Muskelarbeit, Skand.
    Arch. Physiol. 79, 193.
Pitts, G. C., R. E. Johnson and F. C. Consolazio (1944), Work in the heat as
    affected by intake of water, salt and glucose, Amer. J. Physiol. 142, 253.
Robinson, S. (1949), Physiology of heat regulation and the science of clothing,
    ed. L. H. Newburgh (W. B. Saunders Co.), Chap. 5.
Robinson, S., F. R. Meyer, J. L. Newton, C. H. Ts'ao and L. O. Holgersen (1965),
    Relations between sweating, cutaneous blood flow and body temperature in
    work, J. appl. Physiol. 20, 575.
Rowell, L. B., G. L. Brengelmann, J. R. Blackmon and J. A. Murray (1970),
    Redistribution of blood flow during sustained high skin temperature in rest-
    ing man, J. appl. Physiol. 28, 415.
Saltin, B., A. P. Gagge and J. A. J. Stolwijk (1970), Body temperatures and
    sweating during thermal transients caused by exercise, J. appl. Physiol. 28,
    318.
Saltin, B. and L. Hermansen (1966), Esophageal, rectal and muscle temperature
    during exercise, J. appl. Physiol. 21, 1757.
Senay, L. C. (1971), Variations of the ratio cutaneous blood flow: muscle blood
    flow as a determinant in augmentation of plasma protein, Proc. XXV I.U.P.S.
    Congr., München.
Senay, L. C. and M. L. Christensen (1965), Changes in blood plasma during pro-
    gressive dehydration, J. appl. Physiol. 20, 1136.
Snellen, J. W. (1960), External work in level and grade walking on a motor-driven
    treadmill, J. appl. Physiol. 15, 759.
Snellen, J. W. (1966), Mean body temperature and the control of thermal sweating,
    Acta physiol. pharm. neerl. 14, 99.
Snellen, J. W. (1969a), The discrepancy between thermometry and calorimetry dur-
    ing exercise, Pflüg. Arch. 310, 35.
Snellen, J. W. (1969b), Body temperature during exercise, Med. and Sci. in
    Sports 1, 39.
Snellen, J. W., D. Mitchell and M. Busansky (1971), A calorimetric analysis of
    the effect of drinking physiological saline on sweating: an attempt to
    measure mean body temperature, Chamber of Mines of S. Afr. Res. Rep. 14/71.
Stolwijk, J. A. J. and J. D. Hardy (1966), Temperature regulation in man--
    A theoretical study, Pflüg. Arch. 291, 129.
Wyndham, C. H. (1965), Role of skin and of core temperatures in man's temperature
    regulation, J. appl. Physiol. 20, 31.

THE BODY TEMPERATURE "SET-POINT" IN FEVER

K. E. COOPER
Division of Medical Physiology
Faculty of Medicine
The University of Calgary
Calgary 44, Alberta.

INTRODUCTION

That the body temperature regulating mechanism, in man and other animals, behaves as though it were thermostated and that the thermostat is set at an unusually high level during fever has been an accepted descriptive concept for many years. The idea that the central regulating mechanism for body temperature is so reset that it regulates the temperature at a new high level was clearly in the mind of Liebermeister (1875) though he did not actually describe it as such, and of Lefèvre (1911). Lefèvre comments: "La fièvre est donc bien un trouble de la régulation." He quotes experiments of Richet to show that in two febrile subjects cold water immersion produced transitory falls in body temperature which then returned to their febrile levels. He then comments "...ayant chacun dans la fièvre un niveau propre de temperature, comme deux animaux d'espèces différentes". Lefèvre also described the phenomenon of the body temperature appearing to be regulated at a low level for several days following severe fevers. In his classic review Barbour (1921) stated: "in fever it is said that the body thermostat is set at a higher level. A normal temperature becomes apparently interpreted as cold by the temperature centers while one of $40^{\circ}C$ perhaps feels neutral. Similarly the skin nerves seem hypersensitive to cold or hyposensitive to heat whence the subjective chill". Fox and Macpherson (1954), and Macpherson (1959) described observations on a young man in whom the effect of a period of exercise on body temperature and sweat rate had been studied and who then became febrile. They found that the thermoregulatory responses to exercise in the febrile state were closely similar to those which had been observed in the afebrile state, and from this they concluded that the central regulating mechanism in this young man had been reset to operate about a new high norm. Similarly Grimby (1962) showed that the rectal temperature changes during periods of variously graded work were similar in the flush phase of fever to those observed in the normal, healthy state. The difference between his experiments and those of Fox and Macpherson was that the fever in Grimby's subjects was induced by intravenous administration of bacterial pyrogen, whereas in the case of Fox and Macpherson's it was a naturally occurring fever. Cooper, Cranston and Snell (1964) investigated the effect of stimulation of the central warm

receptors on peripheral blood flow both in the afebrile state and during the
plateau phase of fever due to intravenous pyrogen. They found that the dose
response curve - for the stimulation of central warm receptors versus the
peripheral blood flow response - was not significantly different in the plateau
phase of fever from what they had observed in the afebrile condition. Haan and
Albers (1960) noticed in dogs exaggerated responses to cold and reduced responses
to heat during fever and concluded that the   regulator was   intact but
operated about a new and higher level. Some elegant experiments by Belyavsky
(1963) in which they observed thermoregulatory responses to hypothalamic heating,
further support the view that the thermoregulatory system operates about a new
high level in fever. Belyavsky measured the rectal temperature and the
elevations in ear temperature and respiratory rate in response to diathermy
heating of the hypothalamus, thereby raising its temperature by $1 - 2^{o}$C. Intravenous pyrogens caused fever, and the observed responses to hypothalamic heating
were diminished during the rising phase of fever, and they were restored or even
exaggerated during the plateau and defervescence phases.

The concept then seems well supported by experimental evidence but it
describes only the overall response of the central mechanism, and does not
discriminate between various parts of the thermoregulatory equipment of animals,
and the effect which pyrogens may have upon each of these. Also in most of the
work supporting this concept the stimulus used has been one of displacement
towards over-heating. It is more difficult to produce central cooling in animals
during fever and many of the responses may be difficult to observe during such
central cooling, for example further vasoconstriction might be impossible to
observe in the presence of an intense vasoconstriction due to fever itself. Also
much of the work which has been done has been qualitative rather than quantitative. Even in the quantitative work which has been done there has been considerable scatter in results, which would make it difficult to hold the hypothesis
that the central regulator is doing more than regulating roughly in the same way
at about its new level. More precise methods and more detailed analysis of all
thermoregulatory responses during fever are still required for this purpose.
FEVER IN HYPOTHERMIC SUBJECTS

There are some strange instances in which the same hypothesis holds true
but in patients whose temperatures start off at abnormally low levels, Duff and
his colleagues (1961) observed the effect of pyrogens given intravenously to some
patients suffering from spontaneous periodic hypothermia of uncertain origin.
Their patients had low body temperatures in the order of 87°F (30.5^{o}C) and 96°F
(35.5^{o}C). In both instances the characteristic patterns of fever were observed,
with a rise in temperature before the peak of "fever" was reached of approximately the same order as would be expected in patients starting at normal body

temperature. Similar observations were made in patients with chronic hypothermia where thermoregulatory mechanisms were present but operating at an unusually low level by Hockaday, et al. (1962). So it appears that pyrogen induced fever can not only cause a normal body temperature to be raised and the regulating mechanism to behave as though it were set about the new raised "set-point", but a similar reaction can occur in patients who as a result of some intracranial pathology have unusually low, but regulated, starting temperatures. It is also known that in patients with intracranial abnormalities, for example pressure upon the hypothalamus by craniopharyngiomta, an abnormally low central body temperature can be achieved and that in these patients the central warm receptors appear to behave with approximately their usual sensitivity during stimulation (Cooper, 1970). Similarly there is evidence that the central warm receptors can operate at unusually high levels of body temperature with their apparent usual output in terms of vasomotor response. All this adds up to two or more possibilities. First that pyrogens act on some anatomical part of the thermoregulatory apparatus which is itself concerned with the absolute "set-point" about which body temperature is regulated and raise that "set-point" in some way is yet not understood. Second that they act upon cells in afferent or efferent pathways concerned with vasoconstriction and heat production and not on the "set-point" regulator, in such a way as to drive these to an extent dependent upon the local pyrogen concentration. In the steady state the heat loss drives due to the raised central temperature plus the additional experimental loads applied appear to drive the system as though it were regulating about a higher level.

SITES OF PYROGEN ACTION

We must therefore briefly examine the sites of reaction to pyrogens. The first major breakthrough in this field was by Beeson (1948) and Bennett and Beeson (1953) who showed that leucocytes can themselves release a pyrogen. The currently accepted view is that the bacterial pyrogens interact with various tissues, particularly leucocytes, to liberate a new substance which in the case of the leucocyte is a small protein (Murphy, et al, 1971). This gets into the brain to produce the febrile response. The substance is effectively a central nervous system (CNS) poison capable of modifying the response of thermoregulatory cells. It would be interesting to know the relative effects of leucocyte pyrogen on many types of neurone to determine whether the specificity of its sites of action are determined by selective neuronal sensitivity, or by selective permeability of its sites of entry into the brain. There is recent evidence from the results from experiments, in which the cerebral ventricular system of rabbits have been perfused with artificial cerebrospinal fluid during the development of febrile reactions, that leucocyte pyrogen enters the brain via the blood stream and not via secretion by the ependyma and subsequent diffusion through the walls of the third ventricle, (Feldberg, Veale and Cooper

1971). The evidence of microinjection studies (Cooper, Cranston and Honour, 1967,
Jackson 1967, Repin and Kratskin 1967) is that leucocyte pyrogen acts within the
anterior hypothalamus and preoptic region to cause fever. Further evidence
(Rosendorff et al, 1970) has been acquired recently that there may be an additional area in the midbrain which is pyrogen sensitive. While on the surface
these studies seem to make the localization of the sites of action of pyrogen
explicit there is one possible danger in their interpretation. The substance
getting into the brain from the blood stream might not necessarily be the same
as that which is microinjected into the brain in the studies quoted. In other
words some subtle alteration might take place in the leucocyte pyrogen molecule,
during its passage through the endothelium of brain blood vessels and its
diffusion to receptor sites, which would make it a different substance from
that microinjected. So while the cumulative evidence suggests that the microinjected material has very specific sites of action, it is possible that in the
naturally occurring state these might not be representative of the material
which actually reaches the hypothalamus. Such a hypothesis however unlikely
has to be brought to mind.

LOCAL LEUCOCYTE PYROGEN CONCENTRATION

        Now the central action of leucocyte pyrogen on whatever cells it attacks
depends in effect on that quantity of pyrogen reaching the sensitive region.
This will be determined by the amount of leucocyte pyrogen entering the blood
stream and the rate at which it is cleared from the blood. In fact it will be
determined by the time concentration curve for leucocyte pyrogen in the blood.
Then the proportion of the total circulating leucocyte pyrogen present in
arterial blood reaching through the blood vessels of the hypothalamus will be
a determinant of the local concentration within that sensitive part of the brain.
Again leucocyte pyrogen is presumably either destroyed or excreted from the
sensitive part of the brain. The rate of destruction or excretion of leucocyte
pyrogen will be another determinant of the local concentration within the hypothalamus and this in turn will be a determinant of the fever "set-point". It is
also possible that leucocyte pyrogen may release or cause the release of one or
more transmitter substances whose local concentration will be the ultimate determinant of the behaviour of the thermoregulatory system. Their concentration will
be determined by the quantity of leucocyte pyrogen causing their release and the
destruction or excretion of these substances. Some recent evidence has come to
light (Cooper and Veale 1971, unpublished data) that leucocyte pyrogen, or the
febrogenic substances which it causes to be released, may be excreted through the
ependyma of the third ventricle into the cerebrospinal fluid. This statement is
based on the observation that completely filling the ventricular system with an
inert oil, which of itself has no action on the thermoregulatory system,

potentiates and prolongs the effect of intravenous leucocyte pyrogen. In some
instances an intravenous injection of leucocyte pyrogen which had caused fever
of just less than $1^{o}C$ in normal animals was found in animals with oil filled
ventricles to cause fever of higher than $1.5^{o}C$, and lasting up to 48 hours
instead of the usual 1 to 2 hours. Thus the first determinant of the "set-
point" of fever will be the relative quantity of that CNS poison which we call
leucocyte pyrogen getting into and being destroyed or excreted out of the hypothalamic and midbrain areas. It becomes important in many studies to realize
that an intravenous injection of pyrogen produces presumably a very high local
concentration of leucocyte pyrogen in the hypothalamus immediately following
the injection. This will then decay whereas a similar height of fever can be
maintained by a smaller initial intravenous bolt followed by a prolonged
infusion of leucocyte pyrogen infusion. A number of studies therefore on the
response of the hypothalamus, and on antipyretic agents, are better carried out
during the plateau stage of temperature following the intravenous infusion than
they are immediately following a large intravenous bolt of pyrogen.

We must now return to examine a little more closely the sites at which
leucocyte pyrogen might act to produce the apparent change in "set-point"
There have been excellent studies of the action of intravenous pyrogen upon
hypothalamic thermosensitive cells, namely those by Eisenman (1969 and 1970),
Wit and Wang (1968), Cabanac, Stolwijk and Hardy (1968). These authors have
all studied the behaviour of single units recorded in the hypothalamus following
intravenous bacterial pyrogens of one sort or another. Their relevance to the
discussion is of course dependent upon whether the neurones studies are themselves part of the thermoregulatory apparatus. And in view of the recent study
of thermosensitive neurones in the sensorimotor cortex, Barker and Carpenter
(1970) one remains sceptical until a direct correlation between single unit
activity and thermoregulatory responses have been demonstrated clearly in the
unanaesthetized animal. Eisenman's observations were that the response, in
terms of change of frequency with unit change in temperature, of thermosensitive
units from the rabbit hypothalamus was depressed following intravenous typhoid
vaccine. Wit and Wang's results demonstrated that single units recorded in the
hypothalamus had decreased firing rates following pyrogen but that this only
occurred if the body temperature was elevated. At a normal body temperature of
$38^{o}C$ the activity of the recorded cells was unaltered. Cells which were not
temperature sensitive did not show any alteration in their response during the
pyrogen administration. Eisenman recorded from cool sensitive cells, which he
considers to be interneurones connected to some type of cell whose response is
increased by lowering the local temperature, and found that pyrogen increased
their firing rate even at $38^{o}C$. He therefore came up with the view that there

may be a direct action of pyrogen on the central thermodetectors decreasing
their thermosensitivity at above normal temperatures, and causing an increase
in the activity of cool sensitive interneurones. If pyrogen acts on thermosensitive neurones to produce its effect on the apparent "set-point" then it
must do so within the hypothalamus for, in some extremely carefully conducted
experiments reported by Bard, Woods and Bleier 1970, it was demonstrated that
fever was not obtainable in cats with brain stem transactions between the
mesencephalon and the hypothalamus, or between the pons and the mesencephalon.
They present a critical analysis of previous work suggesting that the latter
might be the case, and established conclusively the need for hypothalamic tissue
to be intact in order for intravenous pyrogen to produce fever.

THE ROLE OF MONOAMINES

       The question as to whether pyrogens can act on monoaminergic pathways,
causing fever by the liberation of the appropriate hyperthermic transmitter
substance, continues to be a contentious issue. An observation by Cooper,
Cranston and Honour (1967) that depleting the hypothalamus of monoamines, to an
extent which made them undetectable by formaldehyde fluorescence microscopy, did
not entirely abolish the fever due to intravenous pyrogen would seem to make
these monoaminergic pathways unnecessary for the development of some part at
least of the fever reaction. However more quantitative studies by Teddy (1971)
in which he analysed the mean fever responses to intravenous leucocyte pyrogen in
groups of animals which were either otherwise normal, or which as the result of
locally injected p-chlorophenylalanine (PCPA) or α methyl-p-tyrosine (AMPT) had
considerable depletion of hypothalamic hydroxytryptamine or noradrenaline,
showed that there was some apparent dependence of the febrile response on these
amines. In the rabbit, which normally responds to topical hypothalamic
injection of noradrenaline by rise in body temperature and to 5-HT by a fall in
the body temperature, 5-HT depletion enhanced the febrile response to intravenous pyrogen whereas noradrenaline depletion decreased the response. It may
be then that the monoaminergic pathways are linked to a mechanism which can
modulate some non-aminergic basic fever producing mechanism.

OTHER SUBSTANCES

       The picture has been made even more muddy recently by the appearance of
evidence that other substances may be involved in determining a new "set-point"
in fever. For example Milton and Wendlandt (1970) and Feldberg and Saxena (1971)
had shown that extremely small amounts of prostaglandin E in the cerebrospinal
fluid will raise the body temperature. Myers and Veale (1971), and Feldberg,
Myers and Veale (1970), have drawn attention to the possible role of calcium
ions in the general area of the hypothalamus in determining the level of body
temperature. Further Myers (1971) has shown that he can maintain a monkey's

"set-point" within limits at any level he wants by varying the proportions of
calcium and sodium ions in an artificial tissue fluid perfused through the
posterior hypothalamus. The postulate has been made that fever might be
induced by lowering the tissue fluid concentration of calcium ions with respect
to the concentration of sodium ions in the posterior hypothalamus. Precisely
how leucocyte pyrogen getting into the brain could bring this about is not
clear. It could be that leucocyte pyrogen acts on some cells, possibly
primary thermodetectors, in the anterior hypothalamus and that these release
at terminals around second or third order neurones of the posterior hypothalamus
substances which change the local concentration of calcium. It could be that
the substance which gets into the hypothalamus to cause fever might differ from
that which is injected, as was previously stated in studies attempting to locate
the site of action of leucocyte pyrogen, and that this more refined form of
leucocyte pyrogen could act as a calcium chelating substance in the posterior
hypothalamus. Myers (1971) has produced some evidence that if the hypothalamus
is preloaded with ^{45}Ca and then "push-pull" perfused, the amount of radioactive
calcium appearing in the perfusate decreases during pyrogen induced fever.
However, it is also possible to consider that the changes in pyrogen induced
^{45}Ca changes in "push-pull" perfusate might be related to changes in the blood-
brain permeability for calcium during fever, whereby an exchange between normal
calcium and radiocalcium across the blood-brain barrier would be achieved more
rapidly. It is still possible to think of experimental changes in body temperature "set-point", by altering the ionic constituents in the posterior hypothalamus, as being artifactual; and experimentally determined "set-point"
would thus be rather like pathological conditions which were mentioned earlier
in this paper. The fact that thermoregulation occurs about the new "set-point"
may not be more remarkable than it was in the case of patient whose hypothalamus
was compressed by a tumour.

EFFECTS OF PERIPHERAL SENSORY INPUTS

So far we have only considered the possible effect of pyrogen upon
central thermoregulatory structures and we have not yet considered the possibility that pyrogens might alter the sensory input and by so doing modify the
apparent "set-point" mechanism. The sensory experience at the beginning of
fever, namely the chill, is well known to most people. Not only does the
sufferer feel cold but there is an apparent altered quality in sensory information derived from touching the skin. One of the central effects of leucocyte
pyrogens is to produce vasoconstriction in the skin. Both from microinjection
studies in the rabbit and from studies in patients with various levels of
spinal cord transactions, (Cooper, Cranston and Honour, 1967, Cooper, Johnson
and Spalding (1964) it can be safely stated that the peripheral vasoconstriction

associated with fever is an effect of the pyrogen on the brain and not upon the
peripheral blood vessels, efferent sympathetic nerves, or the isolated spinal cord.
The vasoconstriction of this intense type, of course, leads to a fall in skin
temperature and could therefore exaggerate the information derived from cold
nerve endings in the skin relaying into the brain. It is also possible that
somewhere either in the cord, or in the brain stem the quality of information
reaching thermoregulatory centers from the peripheral temperature receptors
could be modified by pyrogen. Whilst various aspects of the behaviour of
warmth receptors in the skin have been looked at during fever, Bryce, Smith
et al (1959), no quantitative study has yet been made on the cold receptor
activity or its sensory integration during fever. And yet the high sensitivity
to cold afferents in patients experiencing the rigor phase of fever, and the
strange and altered quality of sensation would seem to indicate that such a
study would be worth while. If, however, sensory information from cold receptors
in the skin, altered because of the centrally induced vasoconstriction, were of
great importance in determing the new "set-point" during pyrogen induced fever,
then one might expect an alteration in the height of fever produced by a standard
dose of pyrogen at greatly different environmental temperatures. One study
reported by Bannister (1960) indicated the same order of fever in one subject
to a standard dose of pyrogen at a room temperature of $93^{o}F$ as was observed at
$75^{o}F$. In the two instances the vasoconstriction should have produced greatly
different degrees of skin cooling. In this experiment however it is possible
that the vasoconstriction was not of the same intensity at the two different
temperatures, and as skin blood flow was not measured the results, though highly
suggestive, cannot prove or disprove the importance of skin temperature input in
setting the new level in fever. A preliminary series of experiments has been
done, by Mr. MacLachlan and Mr. Roark and myself in Calgary, in which a group
of rabbits was given a standard dose of leucocyte pyrogen intravenously under
various environmental conditions. The experiments were done at 2°C room temperature, and at 23°C and at 30°C. There was no appreciable difference in the
height of fever (0.7^{o}C rise approximately) at any of the environmental conditions.
Intense vasoconstriction was observed in the ears under all conditions. This
preliminary experiment will have to be repeated during a dose response curve to
pyrogen at the three temperatures, because it is possible that the 0.7^{o}C rise in
temperature, observed with this amount of pyrogen, represents an intersection of
three different dose response curves having greatly different slopes. While
this is unlikely it cannot be ruled out on our present evidence, but the initial
suggestion is that quite markedly different levels of skin temperature during
the vasoconstriction phase of fever do not alter the height of the fever. This
preliminary experiment however raised one awkward point. The rabbit's baseline

temperature was approximately the same at 2°C as it was at 23°C but at an
environmental temperature of 30°C the rabbit allows its temperature to drift up
about a degree higher than at the normal room temperature. The rise in temperature produced by the pyrogen was the same as had been observed with the lower
baseline temperature. We do not know whether the rise in baseline temperature,
at the higher environmental temperature, represents a new "set-point". There
seems no reason to believe this and in fact, it could be that this represents
just a failure of the animal to be able to regulate effectively against the
added heat load and so a fortuitous steady level of temperature is reached
above the animals required "set-point". This being so it is interesting that
the rise in temperature produced by pyrogen was, at the plateau of fever, of
the same order as was that observed with the lower baseline temperature at which
the rabbit was definitely regulating. At the same time from previous work,
particularly that of Murphy (1971), it would appear that 0.7°C rise is well
within a normal dose response curve for rabbits. It might then also be explained
on the idea that this CNS poison, leucocyte pyrogen, gets into the brain and
the time concentration product for its action within the hypothalamus is such
that a relatively fixed amount of extra heat is produced and conserved. Thus
the animal's temperature is raised by approximately a fixed figure each time it
receives an intravenous bolt of pyrogen of this order. In other words what
appears to be a new "set-point" might be fortuitous. Clearly it would be again
better to do this type of experiment using a continuous infusion and testing
for regulation above and below the level of body temperature attained during
the infusion. But even this might not be conclusive, for it might be like the
mechanical pressure on the hypothalamus, that is to say, part of the hypothalamus might be diminished in its function leaving the ordinary input regulatory
channels still capable of functioning at whatever new level of temperature was
determined by the damage done. To call this a new "set-point" would, in a
model theory, be quite valid but one wonders whether the term has any particular value under these circumstances. In fact, is the term "resetting the
central mechanism at a new high level" a splendid example of a piece of jargon
which hinders further inquiry and damps down research, or has it in fact any
real value?

It is interesting to note that there is a considerable body of evidence
to indicate that the response to intracranial injection of monoamines may be
modified by the environmental temperature. If the initial observations on the
lack of effect of environmental temperature on the febrile response to pyrogens
is substantiated, then this might provide additional corroborative evidence
that the monoaminergic pathways are not the primary pathways involved in the
fever reaction.

Another interesting discrepancy is that concerning shivering. Stuart
(1961) has very conclusively shown that the main initiating region of shivering
is in the dorso-medial part of the posterior hypothalamus with modulatory
activity capable of being supplied from the septal region. Again this posterior
hypothalamic region is one which does not appear to be affected by microinjection of leucocyte pyrogen. The drive, of course, provided to this area
for shivering could be secondary to the vasoconstriction, but one would expect
to find shivering only in the cooler environments and not in the warmer ones.
However, some earlier evidence (Cooper, Johnson and Spalding 1964) indicated
that with no sensory input from most of the skin and no vasoconstriction in the
skin, in high paraplegic subjects, shivering did occur in the few innervated
muscles of the neck and head left to the subject. It is clear then there is a
central component of shivering which can be driven by leucocyte pyrogen getting
into the brain. And on Stuart's model and the microinjection studies this
would have to be a septal area of stimulation, possibly removing the inhibitory pathway which normally prevents shivering occurring under ordinary ambient
conditions.

BEHAVIOURAL ASPECTS

The behavioural aspect of the dog to changing environmental temperature
when the animal is afebrile as compared with when it had a pyrogen induced fever
was studied by Cabanac, Duclaux and Gillet (1970). In these experiments the
animal was placed in a chamber and it could either turn on an infrared heater
or turn on cool air from an air conditioning unit. They found that as the
temperature was lowered below 30°C an increase in amount of radiant heat was
applied by the animal to itself and as the temperature was raised above 30°C the
air conditioner was used more often. There was a neutral zone between about
$28-29^{o}$C and 32°C where the animal made little use of either compensatory device.
The febrile animal however used a much more frequent self-stimulation with
radiant heat below 29°C than did the normal animal, and appeared to use the
cooling device only for ambient temperatures which were very hot - well above
40°C. These experiments give a quantitative approach to the commonly experienced human behaviour during sharp fevers, namely that of getting into a warm
bed with a hot water bottle. They could be interpreted as providing yet another
means for the animal to raise its actual temperature up to a new "set-point"
as rapidly as possible, since a wide disparity between the two leads to a
sensation of gross discomfort.

ANTIPYRETICS

No discussion of the "set-point" of body temperature during fever would
be complete without mention of the fact that certain chemicals known an antipyretics revert the body temperature back towards its original level during

fever. It is clear that the antipyretics are not a homogenous class of substances, for lipid-soluble ones such as amidopyrin affect the normal body temperature, (Grundman, 1969), whereas the non-lipid-soluble ones such as salicylate
do not affect normal body temperatures at usual concentrations, (Rosendorff and
Cranston, 1968). The salicylates (Cranston et al 1970) do not effect the fever
due to hypothalamic cooling. They are therefore substances with a specific
affect upon the pyrogen induced reaction and there appear to be a number of
possible ways in which they could restore the "set-point" to the usual level
during fever. There is a strong body of evidence now that they do not act
peripherally either upon the leucocyte pyrogen forming system, or upon peripheral
nerves, or the output from some final common paths of the thermoregulatory
system in the brain stem. They could influence the rate at which leucocyte pyrogen got into the brain, they could accelerate the rate at which it was excreted
from the brain or destroyed within the brain, or they could competitively interfere with its action on the cell surface of thermoregulatory cells. There is
some evidence that they might interfere with the passage of leucocyte pyrogen
into the brain. There is a considerable bulk of evidence that they might competitively act at the neuronal surface with pyrogen which has got into the brain.
Whether one can eliminate either of these possibilities entirely on the present
evidence is doubtful, and we require a pure preparation of leucocyte pyrogen
which can be tagged in order to check the entry limitation theory. It would be
nice to be able to check the other hypothesis at the neuronal level with quantitative methods employing a pure substance which we do not at present possess.

     We have a few preliminary observations of the effect of filling the
third ventricle with inert oil upon the salicylate antipyresis induced during
long infusions of leucocyte pyrogen intravenously in the rabbit. There was
some evidence that oil present in the ventricle may impair the action of intravenous sodium salicylate as an antipyretic. Of course it is possible that if
oil in the ventricle prevents the excretion of leucocyte pyrogen, then the local
concentration within the brain would build up to too high a level for the
salicylate to compete with at the concentration used. Conversely the less
likely hypothesis might be that the normal route of excretion is blocked by oil
in the ventricle and consequently the ability of salicylate to speed up this
excretion is impaired. However unlikely this hypothesis we have to bear it in
mind as a possibility. In favour of a local competitive action are the observations of Wit and Wang (1968) who have shown that the altered firing rates of
single units in the anterior hypothalamus preoptic area is again modified back
toward the normal by the administration of salicylate. And the observations
of Cranston et al (1971) who have shown that injections of salicylate into the
hypothalamus will substantially reduce fever due to prolonged infusion of

leucocyte pyrogen. The only problem with this type of experiment is that by
whatever route salicylate gets into the hypothalamus it can exert an effect on
the permeability of vessels in this area, thereby limiting further access of
leucocyte pyrogen or migration of leucocyte, into the region, and so these
experiments are strongly suggestive but always fail to be conclusive. So the
best that can be said at present is that the salicylate action is central and
it is strictly confined to the action of leucocyte pyrogen. This action may be
a multifactorial one, involving both modifications of the rate of influx and
egress of the pyrogen and modification of its action at cell surfaces.

REFERENCES

Bannister, R.G. (1960). Anhidrosis following intravenous bacterial pyrogen.
Lancet. i. 118-122.

Barbour, H.G. (1921). The heat-regulating mechanism of the body. Physiol.
Rev. $\underline{1}$, 295-326.

Bard, P., Woods, J.W., & Bleier, Ruth. (1970). The effect of cooling, heating
and pyrogen on chronically decerebrate cats, in "Physiological and
Behavioural Temperature Regulation". ed. Hardy, Gagge, Stolwijk.
Thomas, Springfield, Illinois, pp. 519-549.

Barker, J.L., & Carpenter, D.O. (1970). Thermosensitivity of neurones in the
cortex of the cat. Science $\underline{169}$, 597-598.

Beeson, P.B. (1948). Temperature-elevating effect of a substance obtained from
polymorphonuclear leukocytes. J. Clin. Invest., $\underline{27}$, 524.

Belyavsky, E.M. (1965). Analysis of the changes of the thermoregulatory centre
excitability during the development of fever reaction. Path. Physiol. &
Exp. Therapy (USSR) $\underline{2}$, 30-34.

Bennett, I.L. & Beeson, P.B. (1953) Studies on the pathogenesis of fever. II.
Characterization of fever-producing substances from polymorphonuclear
leukocytes and from the fluid of sterile exudates. J. Exp. Med. $\underline{98}$,
493-508.

Bryce-Smith, R., Coles, D.R., Cooper, K.E., Cranston, W.I., & Goodale, F. (1959).
The effect of intravenous pyrogen upon the radiant heat induced
vasodilatation in man. J. Physiol. (Lond). $\underline{145}$, 77-84.

Cabanac, M., Stolwijk, J.A.J., & Hardy, J.D. (1968). Effect of temperature
and pyrogen on single unit activity in the rabbit's brain stem. J. Appl.
Physiol. $\underline{24}$, 645-652.

Cabanac, M., Duclaux, R., & Gillet, A. (1970). Thermoregulation comportementale chez le chien: Effects de la fièvre et de la Thyroxine.
Physiology and Behaviour $\underline{5}$, 697-704.

Cooper, K.E. (1970). Studies of the human central warm receptor, in
"Physiological and Behavioural Temperature Regulation" 224-231. ed.
Hardy, Gagge, & Stolwijk, Thomas, Springfield, Illinois.

Cooper, K.E., Cranston, W.I., & Snell, E.S. (1964). Temperature regulation during fever in man. Clin. Sci. 27, 345-356.

Cooper, K.E., Johnson, R.H., & Spalding, J.M.K. (1964) Thermoregulatory reactions following intravenous pyrogen in a subject with complete transection of the cervical cord. J. Physiol. (Lond). 171, 55-56 P.

Cooper, K.E., Cranston, W.I., & Honour, A.J. (1965). Observations on the site and mode of action of pyrogens in the rabbit brain. J. Physiol. (Lond). 191, 325-337.

Cranston, W.I., Hellon, R.E., Luff, R.H., Rawlins, M.D., & Rosendorff, C. (1970). Observations on the mechanism of salicylate induced antipyresis. J. Physiol. (Lond). 210, 593-600.

Duff, R.S., Farrant, P.C., Levraux, V.M., & Wray, S.M. (1961). Spontaneous Periodic Hypothermia. Quart. J. Med. 30, 329-338.

Eisenman, J.S. (1969). Pyrogen-induced changes in the thermosensibility of septal and preoptic neurones. Amer. J. Physiol. 216, 330-334.

Eisenman, J.S. (1970). The action of bacterial pyrogen on thermosensitive neurones, in "Physiological and Behavioural Temperature Regulation". ed. Hardy, Gagge, & Stolwijk, Thomas, Springfield, Illinois. pp. 507-519.

Feldberg, W.S., & Saxena, P.N. (1971) Fever produced in rabbits and cats by prostaglandin E, injected into the cerebral ventricles. J. Physiol. (Lond). 215, 23-23P.

Feldberg, W.S., Veale, W.L., & Cooper, K.E. (1971). Does leucocyte pyrogen enter the anterior hypothalamus via the cerebrospinal fluid? Proc. XXV Int. Cong. Physiol. Sci. Munich.

Fox, R.H., & Macpherson, R.K. (1954). The Regulation of body temperature during fever. J. Physiol. (Lond). 125, 21 P.

Grimby, G. (1962) Exercise in man during pyrogen-induced fever. Scand. J. Clin. & Lab. Invest. 14, Suppl. 67.

Grundman, M.J. (1969). Studies on the action of antipyretic substances. D. Phil. thesis, University of Oxford, England.

Haan, J., & Albers, C. (1960). Über die Auslösung thermoregulatorischer Reaktionen beim Hund unter Pyrogen-Einwirkung. Pflügers Arch. ges. Physiol. 271, 537-547.

Hockaday, T.D.R., Cranston, W.I., Cooper, K.E., & Mottram, R.F. (1962). Temperature regulation in chronic hypothermia. Lancet. i. 428-432.

Jackson, D.L. (1967). A hypothalamic region responsive to localized injections of pyrogen. J. Neurophysiol. 30, 586-602.

Lefèvre, J. (1911). Chaleur Animale et Bioénergétique. Masson, Paris, p. 354.

Liebermeister, C. von (1875). Handbuch der Pathologie und Therapie des Fiebers Vogel Leipzig.

Macpherson, R.K. (1959). The effect of fever on temperature regulation in man. Clin. Sci. 18, 281.

Milton, A.S. & Wendlandt, S. (1970). A possible role for prostaglandin E, as a modulator for temperature regulation in the central nervous system of the cat. J. Physiol. (Lond). 207, 76 P.

Murphy, P.A., Chesney, P.J., & Wood, W.B. Jr. (1971). Purification of an endogenous pyrogen, with an appendix on assay methods, in "Pyrogens & Fever" ed. Wolstenholme and Birch, Churchill Livingstone, Lond. pp. 59-79

Myers, R.D. (1971). Hypothalamic mechanisms of pyrogen action in the cat and monkey in "Pyrogens and Fever". ed. Wolstenholme and Birch, Churchill Livingstone, Lond. pp. 131-146.

Myers, R.D., & Veale, W.L. (1971). The role of sodium and calcium ions in the hypothalamus in the control of body temperature of the unanaesthetized cat. J. Physiol. Lond. 212, 411-431.

Repin, I.S., & Kratskin, I.L. (1967). On the analysis of hypothalamic mechanism in the pyretic reaction. Fiziol. Zh. USSR. 53, 336-340.

Rosendorff, C., & Cranston, W.I. (1968). Effects of salicylate on human temperature regulation. Clin. Sci. 35, 81-91.

Rosendorff, C., Mooney, J.J. (1971). Central nervous system sites of action of a purified leucocyte pyrogen. Amer. J. Physiol. 220, 597-603.

Stuart, D.G. (1961). Role of the prosencephalon in shivering in "Arctic Biology and Medicine", Transactions of the first symposium. Neural aspects of temperature regulation. Arctic Aeromedical Lab., Fort Wainwright, Alaska, ed. Viereck, pp. 295-399.

Teddy, P.J. (1971). in Pyrogens and Fever, ed. Wolstenholme & Birch, Churchill Livingstone, Lond. pp. 124-127.

Witt, A., & Wang, S.C. (1968). Temperature - sensitive neurones in preoptic/anterior hypothalamic region: action of pyrogen and acetylsalicylate. Amer. J. Physiol. 215, 1160-1169.

JAMES D. HARDY
John B. Pierce Foundation Laboratory and Yale University,
New Haven, Conn. U. S. A.

It is the aim of this summary of the series of essays which constitute the principal chapters of this volume to provide a perspective. The student of biological temperature regulation may have a number of interests which will determine his particular approach to the problems of how living organisms, highly vulnerable to small temperature changes, manage to survive and flourish. Thus, it is that the biologist, the psychologist, the engineer and the physician share an interest in biological temperature regulation but from widely different points of view. One area which provides a meeting of these interests seems to be that of modelling the temperature regulator and thereby all the essays will be indirectly reviewed by a short summary dealing with models of temperature regulation. Perhaps also the student may have other questions. For example, do we have enough models? Can we use to advantage those already developed? Have models assisted in understanding biological temperature regulation and what are the prospects of their continued development?

More than 100 years have elapsed since Claude Bernard (1865) stressed the importance of regulation of the "milieu interieur" and it was almost 200 years ago that Lavoisier explained the source of animal heat as due to oxidation of food in the body. The generations of physiologists since 1780, who have studied the "fever problem" and the effects of temperature on living organisms include those giants of our science who set our experimental foundations and our theoretical goals by contributing a model of temperature regulation of one sort or another. Starting somewhat arbitrarily, with the 1880-1890 decade, within which was discovered the importance of the brain in control of body temperature, I counted the number of models offered each decade and the results of this brief survey are indicated in figure 1. Such a compilation requires some specification of what is meant by a "model" and since I have used Grodins' (1970) concept of a scientific "model" the term is broad. The data indicate that in the early years only one or two models were offered in a decade but that in the 1950's interest in models began to increase and in the ten years 1960-1970 more than 20 models were published to explain a concept of some aspect of temperature regulation. Plotted for comparison is the total number of papers using the word "model" in the title and appearing in selected years between 1950 and 1970. In 1965 - '70 there were more than 1500 such papers, at least indicating a growing popularity of the word. I should mention that the slopes of these lines are somewhat greater than the rate of growth of the shelf space required each year for Biological Abstracts. Perhaps one could infer that modelling in the temperature regulation field is characteristic of biological research in general and that the usefulness of modelling techniques is probably on the increase, so that the 1970's may provide as many as 50-100 different models of temperature regulation.

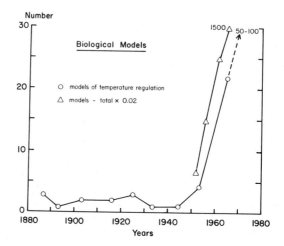

Fig. 1 - Number of models of temperature regulation announced each decade since
1880 (o). Number of papers in selected years between 1950 and 1970 with
word "model" in title (∆).

Many of us have been using modelling for some years but have given little
thought to the general problem of what constitutes a biological model. Professor Grodins in his paper before the 1968 New Haven Symposium outlined for us
in some detail what the biologists mean by modelling and how they use them. I
have slightly modified Professor Grodins' charts to focus a little more on biological temperature regulation and fig. 2 shows in its general form his ideas
but expanded somewhat to illustrate the particular types of models developed for
thermoregulation.

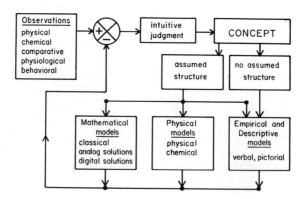

Fig. 2 - The science "servosystem" and the process of developing and utilizing
models of temperature regulation.

The beginning and end of the process is our observation of the real world, as related to temperature and biology. Measurements are made in many areas and the results are turned over in our minds by a process which we do not understand. I have called it "intuitive judgment". From this personal manipulation of the data a concept develops which one may like well enough to offer to his colleagues as a synthesis or explanation of his observations. As indicated in the lower boxes, our descriptive synthesis or "model" will depend largely on the personal preference of the investigator and how comfortable he may feel with is assumptions. There is a long tradition in biology to avoid assuming anything, if possible, and the conservative minded scientist may thus express his ideas in empirical and descriptive arrays verbally or perhaps pictorially. At some later stage he may be inspired to make a mechanical, electrical or even chemical model (for example, Lilly's iron wire model of nerve conduction) for use in teaching. For such models the author must commit himself to some assumptions as to the structure or internal arrangements to explain his process. In the past, many of these analog models have been built and have been of great use in furthering understanding and teaching. The most difficult models for investigators of thermoregulation are the mathematical models and the reason seems to be the general limitations of classical mathematics to solve the nonlinear equations which the author must write in explanation of his data rather than the lack of familiarity with mathematical methods. With the advent of computers, at reasonable cost, this problem has been overcome and there now exist several mathematical models that give reasonable explanations of much that is known regarding the physiological responses of man to heat, cold and exercise. Figure 2 indicates, however, that the purpose of any and all forms of models is to make a comparison of the "concept" with the real world so that if the investigator stops with his "model", much of its usefulness is lost no matter how well conceived and executed it may be. I have shown the model as serving as a negative feedback in our scientific process which inhibits further intuitive judgments and concepts. In some instances this inhibition can completely stifle the experimental effort because an author may feel he has the answers. Fortunately, the process is more likely to stimulate experimental effort and the development of new data and thus, maintain the processes of interplay between the facts of the real world and the imagination of physiologists. For best results, the model maker must keep his scientific servosystem in operation using all parts of it continually; that is, by continually doing experiments to challenge the model and by building new models as dictated by the data. This process may involve performing experiments to challenge the models of our colleagues or the somewhat less pleasurable consideration of new data which may throw doubt on our own cherished concepts.

Having outlined what is meant by a "model" and how it serves to help in the communication between physiologists, teaching of students and in furthering research let us try to answer the questions; where, when and what kinds of models have been developed for biological temperature regulation? Using the scheme in fig. 2, I will attempt a three dimensional review by recalling some of the typical models of the past as examples to provide some answers to these questions.

Verbal Models

These models were the first to be used, and possibly Claude Bernard, with his intense preoccupation with biological regulation, could be said to have offered the first of these models by his famous statement concerning the "fixité du milieu interieur". Certainly, the source and constancy of animal heat were known and speculated upon long before the 19th century and thermometry in physiology appeared early in the 17th century. Between 1880 and 1890 the brain was identified as important in temperature regulation by Richet in France (1885) and Ott (1887) in the U.S. Using a slender probe these authors demonstrated that if a puncture was made deep into some areas in the base of the brain of a homoiotherm (cat, dog, rabbit, etc.) a high fever developed. In his physiology textbook (1904) Ott states - "When a normal animal is subject to heat or cold it regulates its temperature and keeps it at a fixed point". He also identified "Thermotaxic Centers. - These centers compose the thermogenic, thermo-inhibitory, and thermolytic centers, as the aim of all is to regulate the temperature". Ott's remarks have quite a modern ring even though they seem to have had little impact on his colleagues of the 19th century. So far as I am aware, Ott made no pictures or diagrams of his "concept" but his statements seem clear enough.

Another model, which had the advantage of using more quantitative data, was that proposed by Rubner (1902), who identified the ranges of ambient temperature evoking "Chemical Regulation", "Physical Regulation" and "Rise of Metabolism". These concepts continue to be useful today and the terms "physical" and "chemical" regulation are in current use. In terms of our scheme, it seems that Rubner's "Concept" was expressed as both a verbal and pictorial model which physiologists found useful in classifying their observations of thermoregulatory responses to environmental temperature. It would be unfair to leave the work of Rubner without comment on his major contribution - establishing the applicability of the first law of thermodynamics for animal metabolism. After Rubner there were few serious discussions of the law of conservation of energy as applying to biological systems. Even today, when physiologists are studying the energy ba-lance during "negative" work, they begin with the first law and base their hypotheses on the fact that energy may change form but none is lost or created in the process of "negative" work.

An important verbal model which must be referred to was proposed by Hans Meyer in 1913. He stated that "there are two sets of antagonistic functions which overcome and work in opposition ... The entire apparatus can be thought of as ... two local and perhaps entirely separated but integrated centers: a thermogenic and therefore heat conserving ... and a thermolytic or heat loss and temperature lowering center. We will designate these as warm and cold centers." He continued with the interesting remark which some of us might question - "The foregoing is not new and little has come from the concept". Perhaps we would agree that he was restating much of Ott's concept but Meyer's model was more clearly stated and contains much of modern thinking. Bazett in 1949 and Benzinger in 1959 affirmed their belief in Meyer's general concept and many generations of medical students have been taught the Ott-Meyer verbal model as the best available explanation of temperature regulation in man. A principal focus of interest in thermoregulatory research has always been fever and its management, especially in the period before the identification of pyrogenic substances and the advent of antipyretics. Liebermeister in 1871 gave quite a clear statement of his concept of fever: - "The action of a pyrogenic substance must be somewhat as follows: it must by direct or indirect action influence the center of heat regulation so that it regulates at a higher temperature". He also pointed out that "in a fever illness heat regulation is by no means lost but acts in the same way as in health". Is this not the belief in 1971? It could also be allowed that Liebermeister made the first statement of what is now known as the "adjustable setpoint" concept of temperature regulation. In brief, verbal models have been of service to everyone who has been interested in the problem of biological temperature regulation and very likely will continue to be essential in the transition from "intuition" to "concept" regardless of the final form of the model development. However, a major problem with the verbal model is, how does one formulate an adequate challenge to it? The pro-cess of "ruling out" all other possible explanations is a difficult and per-

haps impossible task for so complicated a system as thermoregulation. Perhaps
it was for this reason that Hans Meyer gave voice to his frustration by saying
"little has come from the concept". That is, our research mill stops because
there is no clear way of predicting hypothetical results with which data from
experiments can be compared. Another step in model making must be taken to
state more exactly what is meant by the verbal statement of the concept. His-
torically, the next step was an immediate jump to the mathematical model proposed by Professor Alan C. Burton in 1934, which I will discuss briefly below.

    Before leaving the subject of verbal models mention should be made of a
broad topic - why temperature regulation? In the first essay of this volume,
Dr. Dawson presents a conceptual model involving all the strategies available
to living forms to deal with the thermal threats to life. Both behavioral
and physiological temperature regulation are considered as they are combined
by various species to determine the particular strategy for meeting the dangers
of environmental heat and cold, survival in the face of numerous predators and
the search for food. He points out that evolution has produced animals with
increasingly higher internal body temperatures and that survival has been se-
cured at the price of higher resting energy consumption. Temperature regulation against environmental cold and overcooling of the body has been of prime
concern. This development has led to increasing risk of becoming overheated
with consequent irreversible denaturation of vital proteins at temperatures
near 45°C. Although many forms have developed physiological responses which
tend to prevent overheating by panting, sweating, etc., others must accept
storage of body heat and a dangerous hyperthermia (44-47°C) during periods of
combat or flight. Behavior can do little to rid the body of the very large
heat loads resulting from high levels of exercise developed during such periods
of stress and the animal then must depend mainly on the physiological mechanisms of temperature regulation for survival. Dawson's verbal model puts behavior and physiology in perspective as they serve in the development of
strategies for meeting one of the most important ecological stresses.

Pictorial Models

    As noted in figure 2, there are two channels indicated between the "concept"
and the "empirical and descriptive models". In one case no structure is assumed and in the other the author commits himself to an arrangement of some
kind between components and processes. This dichotomy becomes particularly
clear for pictorial models. For example, if the diagram is a simple rectangle
representing a "black box" approach no structure need be assumed. However, if
the pictorial model represents several identified components or relationships
between processes, a structure of some sort is assumed and to many physiologists this case will be more interesting. For purposes of teaching no model
is more effective as a lecture device than the pictorial model. The general
purpose of a pictorial model is to bring together the conceptual and biological
components in an easily comprehensible form. The concept behind a pictorial
model is often not contained in the diagram, but arrangements of many components of the system are usually indicated so that anatomical, physiological and
other data can be used to challenge the correctness of the model. Most of the
models of temperature regulation are of this sort, if we include diagrammatic
anatomical sketches and block diagrams under the heading of pictorial models.

    The first of these models (Figure 3) is a diagram of the neurophysiological
components of the thermoregulatory system, and its connection to some of the
control elements, sweat gland, cutaneous blood vessel and skeletal muscle.
According to a later statement by Dr. DuBois, this diagram had great appeal

and many requests were received from colleagues for permission to reproduce it.
The state of knowledge today of the neurophysiology of temperature regulation
extends beyond this diagram, but nevertheless it contains much of what is known.
A more detailed pictorial model, shown in figure 4, was proposed in 1952 by

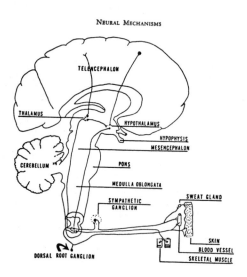

Fig. 3 - Relationship of the nervous system to the control of body temperature.
        (DuBois, 1948)

Hensel which related many physiologically identifiable functions and processes
via information channels of two sorts - stimulation and inhibition. Some question marks indicated areas of the author's uncertainty! The central components
can be identified as responding to blood temperature with the development of two
kinds of signals - one inhibiting (dashed lines) and the other stimulating (solid
lines). Integration of temperature signals from skin and center by two separate
centers is illustrated with command signals for panting, sweating, shivering and
vasomotor control indicated by separate, solid and dashed lines. On first in-
specting this carefully thought out diagram, my immediate interest was focussed
on the function of the black box in the "Center" - how could it inhibit (-) and
stimulate (+) at the same time without two central receptors (warm and cold) or
some functional setpoint to divide the hypothermic from the hyperthermic responses? That is, could such a system really regulate body temperature? The
Hensel model was a distinct departure from the Ott-Meyer concept in that it outlined a dual function for the brain stem centers, i.e. temperature sensing and
signal integration. This model stimulated research activity in at least two directions. Studies were begun to determine experimentally the nature of the sensors in the center and the mode of integration of temperature information. Also,
there followed almost at once attempts to use systems analysis to answer the ques-
tion-could Hensel's model really regulate? It is possible this model provoked more

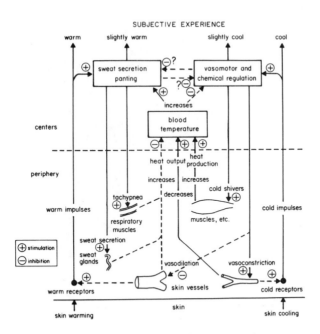

SUBJECTIVE EXPERIENCE

Figure 4 - Diagram of temperature regulating system (redrawn from Hensel, 1952).

various types of investigation than previous models and in this sense the scientific servosystem was accelerated.

The verbal-pictorial model which evoked the most emotionally charged activity was proposed by Benzinger in 1959. The simple diagram proposed by Dr. Benzinger is shown in fig. 5, and has wide coverage both in the scientific and semi scientific press. The model accepts the Hensel idea of a sensing function for the brain stem limited to "warm reception" by a "thermal eye" and limits the thermoregulatory function of the skin to cold sensors analogous to those previously found by Hensel in the nose of the cat. As indicated in figure 5, sweating and vasodilatation are controlled from the central sensor of warmth, A. The shivering response to cold is driven entirely by cold sensors in the skin via "a synaptic, temperature insensitive center, P", in the posterior hypothalamus. Benzinger provides for an inhibitory pathway from the warm sensor to "P", and similarly, the sweating response is inhibited by cutaneous cold sensors. Benzinger also proposed that physiological temperature regulation in exercise and at rest were explainable only with his model. Perhaps it was Dr. Benzinger's stimulating manner of speaking and writing about his model that so influenced his colleagues, but it seems clear that his model provoked challenges from many laboratories over

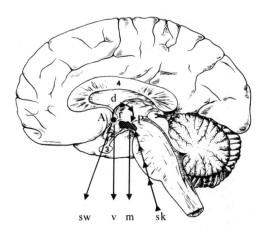

Fig. 5. - Model of temperature regulation in man. (Benzinger, 1963).

the world and the scientific mill was accelerated by his contributions.

Another verbal-pictorial model (Fig. 6) which assumes little structure but has stimulated thought was proposed in 1965 by Chatonnet and Cabanac. This model was the first to propose that behavioral and physiological temperature regulation have common origins. The authors were outlining hypotheses suggesting that the motivation for behavioral thermoregulation originates in feelings

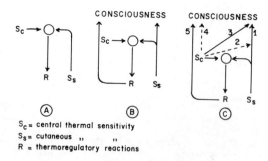

Fig. 6. - Diagrams representing conceptual models of thermoregulation and the interaction of thermal stimuli in producing sensations of warmth, cold and thermal discomfort. (Chatonnet and Cabanac, 1965)

of discomfort related to central (S_C) and peripheral (S_S) thermal stimulation and also to the physiological responses (R). Their diagram, which is deceptively simple, requires the verbal statement of the several hypotheses involved. This model has been stimulating considerable research in behavioral temperature regulation and has been elaborated upon by several authors (Adair, 1972, Hardy, 1971a). Again we may say that our understanding was moved forward by these authors and their model.

Analog Models

The next step in model development is the making of some apparatus or process which represents the concept as well as possible. There have been a number of these to represent the circulatory system; for example, a glass and rubber model of the cardiovascular system was used in our physiology laboratory for teaching medical students. For temperature regulation there have been a number of models used for representing the heat flow patterns from human skin. Among these are the katathermometer (wet and dry) introduced by Leonard Hill (1923), the eupathoscope and thermointegrator used at the Pierce Laboratory in the 1930's, and the Vernon black globe (1932) used to represent the radiation exchange between man and his environment. The idea of these models was to represent in some convenient form this thermal interchange, so that an index of physiological thermal stress could be obtained without necessarily having to expose the man to the particular conditions. These models were not truly thermoregulatory in the sense that they attempted to represent the physiological system, but they were looking at man as a black box and relating his reactions while working in the heat to instrument readings via an "index".

Omitting reference to many early efforts in constructing physical analogs of thermoregulation, I will use for illustration only two, a hydrodynamic model developed by Aschoff (1958) for the study of counter current heat exchange, and an electrical model proposed by MacDonald and Wyndham (1950). Aschoff's model is illustrated in fig. 7. The "extremity", within which the countercurrent heat exchange occurred, was a glass tube with "artery" and "vein" surrounded with

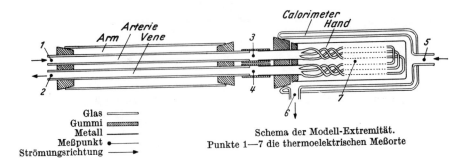

Glas
Gummi
Metall
Meßpunkt
Strömungsrichtung

Schema der Modell-Extremität.
Punkte 1—7 die thermoelektrischen Meßorte

Fig. 7 - Physical model to represent the countercurrent heat exchange (Aschoff and Wever, 1958).

cellulose in water. Flow of water through the system was controlled and measured and the heat transfer into the model hand was measured calorimetrically,
as well as by the rate of perfusion of the hand and the temperature difference
between arterial and venous streams. With the model, Aschoff was able to demonstrate the counter current effect even though the model was at best a rather
rough approximation of the actual physiology of heat transfer by the blood in
the arm and hand. Aschoff's model indicates the importance of using models to
study a subsystem or small part of the entire thermoregulatory system.

    A physical analog of much interest is the electrical circuit proposed origin-
ally by MacDonald and Wyndham in 1950. Using electrical potential as the analog
of temperature and electrical current to represent heat flow, these authors
showed the usefulness of this type of analysis to give added understanding to
experimental data obtained on men exercising in the heat. Their diagram is
shown in figure 8 and includes a representation of a complete control system
with central and peripheral inputs to the closed loops for vasomotor and sweat
rate control. However, they did not attempt to solve the equations for the

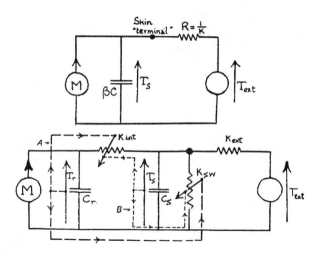

Fig. 8 - Electrical model of thermoregulation (MacDonald and Wyndham, 1950).
        Upper circuit; heat transfer diagram: Lower diagram; a two compartment
        model with thermoregulatory controls for vasomotor and sweating
        functions.

control of their two compartment system but pointed out many of the qualitative
and semiquantitative similarities between the physiological data and the deductions made from the electrical system. The later studies using analog computers are direct descendants of this early simulation analysis. It thus,
appears that the physical analogs have stimulated activity in our research mill,
even though the models which have been proposed are poor representations of our
biological system. They are clumsy to use and experiments with them are done
usually in "real" time.

Biological analogs were not included in Fig. 2 with physical and chemical analogs because such models belong to the real world of physiology and observations on them develop concepts for modelling their own activity. It is often useful to make inferences about a particular organism (man, for example) from studies on other forms and the limitations in the use of biological models are well known.

Mathematical Models

As indicated by Dr. Mitchell in an earlier essay, several advantages are realized if the model maker can symbolize his concepts in mathematical form. The rules of logic for mathematics, having been worked out and accepted, are more understandable and have been outlined in great detail over centuries of use. It is only necessary for the physiologist to order his system in his own mind so that he can put his model into a form which can be described in mathematical terms. In making this simplification much of the realism of the physiological system may be lost just as in the case of physical and chemical analogs. However, the mathematical model can be widely understood, since assumptions must be written in specific, mathematical language, and solutions are obtained using a wholly acceptable logic. Further, the mathematical model is quite flexible in its statement, so that changes can be introduced and results of tests can be obtained rather quickly when compared to physical models. An added advantage is the form of the results produced by a mathematical model; these can be a time sequence, such as obtained from physiological experiment, or as deductions of more complicated relationships between measured variables.

The first of the mathematical models, as has been mentioned, was proposed in 1934 by A. C. Burton. In order to carry through his solution with classical mathematics he assumed the body to be a cylinder with uniform properties throughout; that is, uniform density, specific heat, thermal conductivity and heat production per unit volume. His solution was in the form of Bessel functions for his specified values of axial temperature and skin temperature. He proposed to study the temperature distribution within his model assuming the man was first immersed in a tub of well stirred water at 32.5°C and then the water was rapidly heated to 36°C. The problem was then to determine the temperatures within the body during the warming process assuming that the body rapidly vasodilated, so as to increase the effective conductivity by four fold. The results of the calculations are shown in figure 9, as a series of temperature plots versus time and distance from the center axis of the cylinder. The predictions are quite interesting and indicate:-

(a) that quite a long time will be required (some hours) before complete adjustment;

(b) the temperature of the deep tissues will first decrease before increasing;

(c) the initial and final temperature profiles within the tissues are parabolic.

As Professor Burton pointed out "The correspondence (with physiological data) is satisfactory considering ...". There was no attempt made to follow up on this model and no one has challenged it in any way. Thus, although Professor Burton's model opened the door to a new development in thermoregulatory research, it failed to stimulate additional experiments at the time.

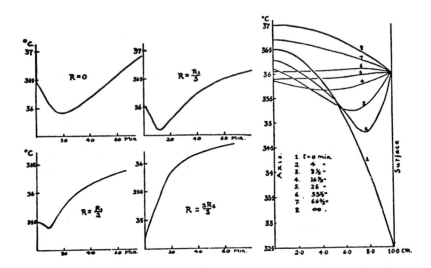

Fig. 9 - Analysis of temperature changes within the body by classical mathematics (Burton, 1934). R = radius of uniform cylinder representing body.
Left: body temperature at center (R = 0) and other distances from axis
following transfer of cylinder from bath at 32.5°C to 36°C. Right:
thermal gradients in cylinder at different times.

In 1961, about thirty years after Dr. Burton's model, a very carefully conceived representation of the heat transfer in man was presented by Professor
Wissler. A diagrammatic representation of this model is shown in fig. 10. Instead of a single cylindrical compartment, the Wissler model is arranged to
roughly represent the human form with heat, extremities and trunk. Furthermore,
each of the six cylindrical elements is composed of an inner core, muscle, fat
and skin layers through which heat is considered to flow radially. Heat transfer
via the blood circulation connects the trunk core with all of the other shells
through arterial and venous channels which allow for counter current heat exchange. The definition of the model has been increased recently by Professor
Wissler, so that it consists of 250 compartments. As a representation of the
passive or controlled system for humans this model is the best that has been
proposed but it requires a rather elaborate computer to handle the computations.
However, it is the only model which predicts realistic values of heat exchanges
and thermal gradients within the body for exposure to cold and for transient
conditions. Dr. Wissler notes that his earlier 1963 model with 15 compartments
required 10 minutes of computer time to produce a solution which is probably as
fast as Dr. Burton could look up the Bessel functions. However, using such a
model is slow and expensive, particularly, if one wishes to add the controlling
system so as to have a complete model of thermoregulation. This model has not
been challenged nor has it as yet stimulated new physiological research.

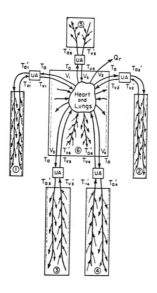

Fig. 10 - Schematic representation of heat transfer in man via tissue
conduction and blood stream for a six cylinder 15 compartment
system. (Wissler, 1961)

Analog Models of the Closed Loop Systems

When the first attempts were made to offer a complete model of thermoregulation about ten years ago, two models appeared at about the same time. Both of
these models included a simplified representation of the passive thermal system
for man but proposed two different theories for the control action of the thermoregulator. The first of these was the result of a collaboration of an engineer and a physiologist, who were studying the effects of ambient temperature,
vapor pressure and air velocity upon subjects working at different rates. Smith
and James (1964) were using heart rate as the index of thermal strain and thus,
were interested in having their model predict the heart rate for alternate
periods of work and rest. For their passive model they used three cylinders to
represent the trunk, i.e. core, overlying fatty layer, and an outer layer for
the skin. The original form of their concept of the regulating system is shown
in figure 11. Although a rather elaborate 6 cylinder model was proposed, they
used only the trunk for their calculations because of the expense of computer
time in calculating heat exchange for a 15-19 compartment system. They made
allowance for heat exchange by counter current flow and otherwise used standard
equations for the heat balance of each of their compartments.

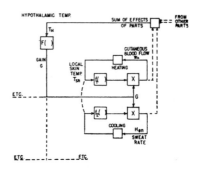

Fig. 11 - Diagram representing control system for human thermoregulation
        (Smith and James, 1964).

The form of their regulator is interesting in that they proposed local skin
temperature control of both the cutaneous blood flow and sweat rate through two
positive feedback loops following a suggestion made by Bazett in 1949. However,
they introduced the hypothalamic temperature as a multiplier which determined
the gain of the two positive feedbacks. From the data available in the literature they selected as best they could values for the local and central gain
functions, F ( ), f1 ( ), f2 ( ), and chose 37°C as the central setpoint temperature; below 37°C the central gain was zero - these two local effects were
inhibited. Finally, they selected relationships between heart rate and the
cardiac output needed for muscle and cutaneous blood flow. As a test of their
model they made a best fit of model predictions to the mean values of heart
rate as measured on four subjects walking on a treadmill at 32°C. As illustrated
in figure 12, the agreement between prediction and experiment leaves little to be
desired. Fortunately, this model was a challenge to a model proposed from our
laboratory (Crosbie et al 1961). The passive system was termed the "slab" model
and was essentially the same* as that used by Smith and James.

*Whether or not one uses linear or cylindrical geometry to describe the complex
human form is a matter of personal preference regarding certain factors such as
convection coefficients; should they appear as parts of the model description
or be retained as environmental factors?

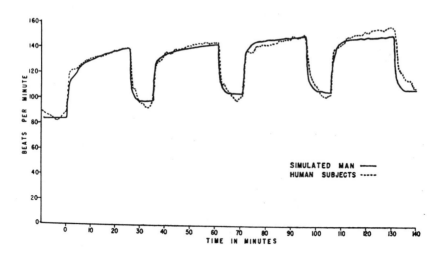

Fig. 12 - Comparison of experimental values of heart rate of four subjects during
alternate work-rest periods with predictions from model. (Smith and
James 1964)

The main difference between the models was the concept of regulation, our proposal assumed all regulation to be of central origin with the skin and muscle
temperatures summing their signals with the hypothalamic temperature signals,
to control cutaneous blood flow, sweating and shivering. For this hypothesis
three setpoints were required, one for the hypothalamus (or core), one for the
muscle layer and one for the skin; no provision was made for a local effect of
skin temperature. The model was tested against steady state calorimetric data
on the responses of nude men to ambient temperatures 22-35°C (Hardy and DuBois
1938). The results are illustrated in figure 13 and it is seen that there is
agreement between model predictions and experimental data (shown as open circles).
Two versions of the effector system are indicated, one of which (dashed line) assumed a limit to the vasomotor control (ΔK). It should perhaps be noted that the
concept of the regulator for this model is one form of what Professor Hammel
some years later called "the adjustable setpoint model".

Thus, the situation faced by the model makers in the mid 1960's was that of
two different concepts of the regulation of body temperature both capable of explaining different sets of experimental data. The summation model was challenged again some years later by a central multiplication model, proposed by
Stolwijk and Hardy (1966) to explain data obtained on subjects who were exposed
transiently to hot and cold environments. At the present time, there are three
models, developed by analog computer methods, awaiting data to exclude one or
all of them; namely,

1) the Smith and James - central hypothalamic signal gain control of local
   skin temperature effector loops;

2) the Crosbie-Hardy model - regulation achieved by summation of signals
   from hypothalamus (or core) deep and superficial receptors (no local
   skin temperature effect);

3) the Stolwijk-Hardy model - regulation by gain control through signals from skin and hypothalamus (no local skin temperature effect).

There are modifications of these models but the above concepts are the ones which have been tested against data. However, those efforts have not been extensive enough. For example, none of the models does an acceptable job of predicting human responses to cold environments. On the other hand, they appear to have stimulated considerable research both on man and animals, so that they seem to be furnishing food for the research servosystem even if they do not seem to be as definitive, as was once hoped, in deciding which of several concepts is nearest correct. Physiological data will have to serve this function much as it has in the past.

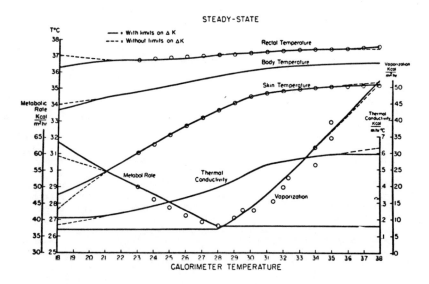

Fig. 13 - Comparison of experimental data from calorimetric studies of men exposed in nude, resting state to temperatures 22-35°C (open circles) with computer predictions (solid lines). Heat loss assumed equal to metabolic rate. (Crosbie et al 1961).

One application of the analog model should be mentioned, since it is likely to be referred to in engineering circles. Models are now being used to predict the effects of thermal environments on man during space travel, lunar exploration and in working conditions of high temperature and water vapor pressure. The Stolwijk modification of his earlier model described in 1966, is being extensively used by the U. S. Space Agency in simulations of space environments and Fanger (1967) has used an empirical model to predict the effects of the thermal environment on feelings of discomfort. Also, Gagge (1971) has introduced for the air-conditioning engineers, a simplified thermoregulatory model (a two layered cylinder) to predict not only conditions compatible with thermal comfort but the effects of clothing, work and the limits of human tolerance to heat exposure. Gagge uses linear geometry for his passive system and adopts concept 3, as best representing the controlling system of the resting man. For conditions of work, he uses a combination of concepts 1 and 2 following the recent Stolwijk model of the controlling system. Since this concept includes the most recent physiological data, it should perhaps be stated as a 4th concept as follows:-

  4) the Stolwijk model - regulation achieved by central signal (90% hypothalamic plus 10% skin) multiplied by local temperature action on sweat glands and cutaneous blood vessels.

The analog model recently proposed by Wyndham and Atkins (1968) is of the same general form as the latest Stolwijk model in that they included the effects of local skin temperature on sweat rate and peripheral conductance.

The use of these analog models to predict all of the various human responses to temperature may be somewhat premature because none of the models now available is completely satisfactory. However, the need for a good model is obvious as a practical matter to assist the engineers and applied physiologists. Perhaps more stress should be put on the recognized limitations of our various models lest some be lead to the conclusion that the problems of biological thermoregulation have been solved. Fortunately, for the viability of research in the area, newer varieties of models are being proposed which indicate more of the complexity of the problem and I shall close my discussion with mention of two of these which have appeared recently.

Neuronal Models

Although since 1885, it has been generally accepted that brain cells are some-how involved in physiological regulation, it is only since 1960 that physiologists have begun to think in any detail about the interneuronal relationships that may be involved. It was in 1961, that temperature sensitive units were reported in the preoptic region (Nakayama et al. 1961) and also by this time data on the responses of a variety of animals to selective heating and cooling of special areas such as the spinal cord (Thauer, 1967), hypothalamus (Hammel et al, 1960), scrotum (Waites, 1961), etc. were being made available. Pickering (1958) and his colleagues had already clearly outlined the thermoregulatory responses in man to pyrogens and to local heating of hands and feet of normal and para-plegic men, so that by 1960 there was a need to attempt formulation of a "con-cept". As yet, the few neuronal models which have been attempted are restricted to limited parts of the system and none has attempted an overall model of the neural component of the thermoregulating system. Perhaps for this reason, there has been no advance of the neuronal models beyond their present pictorial state.

One of the recent neuronal models to be formally proposed is that of Professor H. T. Hammel (1965). This model, based on the few data available on preoptic unit activity, was principally aimed at explaining physiological experiments in much the same way as Hensel's model of 1952. The Hammel concept of "adjustable setpoint" is illustrated in an earlier essay in this volume.

Hammel's model departs sharply from the Liebermeister, Meyer, Bazett concept by indicating, not two "centers" acting reciprocally (one for heat conservation and the other for heat loss) but rather proposes that the peripheral afferents modify the central error signal by adjusting the "T_{set}" of the brain stem so as

to produce the appropriate command or effector outputs. This model has stimulated research both in neurophysiology and thermalphysiology and perhaps it is
fair to say that it is today a focal point of much controversy. It appears that
the model has wide appeal particularly among biologists and "behaviorial" physiologists. Support also comes from neurophysiologists studying the interaction
of peripheral thermal stimulation and single units of the hypothalamus (see
Hellon's essay). However, there are others who are reserved on Hammel's concept
of the "adjustable setpoint" even though endorsing Professor Hammel's departure
from the Liebermeister model.

   A second neuronal model, proposed from my laboratory, is based entirely on the
data from experiments on single unit activity in response to local and peripheral
temperature changes. The aim of the model, illustrated in figures 14a and 14b
was to analyze the responses of all of the observed types of units which have
been reported so far and to see if they could be fitted into a network of some
sort. Although units of the brain stem respond to a wide variety of stimuli
(visual, auditory, mechanical, noxious, chemical as well as temperature) the
question posed was limited to what logical relationship could be visualized between the various types of units already identified on the basis of their response to temperature changes in the central nervous system. A more complete
model would include units responding to skin temperature (Hellon, 1971). Assumptions were necessary regarding transmitter substances, at least two were
required. All units reported could be fitted into some place in the network and
outputs of the cold and warm networks seem to be logical (but somewhat redundant)
from a thermoregulatory viewpoint. The networks as conceived support some aspects of the models of Bligh, Hammel and Cooper (note essays in this volume) and
indicate that many thermal afferents enter the preoptic-anterior hypothalamic
area from various body locations to be summed. However, the circuits indicate
that the preoptic sensors of temperature are not affected by peripheral thermal

PRE-OPTIC-ANTERIOR HYPOTHALAMIC INTERNEURONAL NETWORK
(WARM)

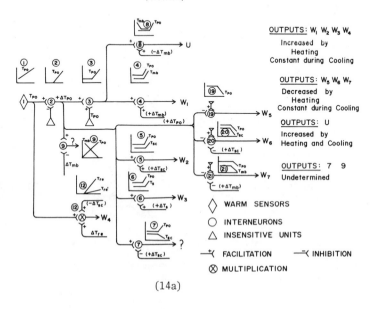

(14a)

PREOPTIC – ANTERIOR HYPOTHALAMIC INTERNEURONAL NETWORK
(COLD)

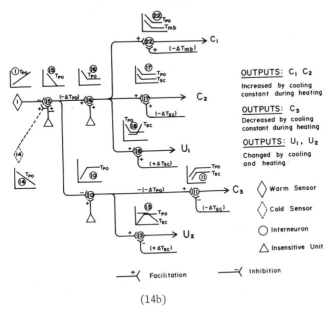

OUTPUTS: C_1 C_2

Increased by cooling
constant during heating

OUTPUTS: C_3

Decreased by cooling
constant during heating

OUTPUTS: U_1, U_2

Changed by cooling
and heating

◇ Warm Sensor

◇ Cold Sensor

◯ Interneuron

△ Insensitive Unit

—⁺⟨ Facilitation   —⟨ Inhibition

(14b)

Fig. 14 - Neuronal model based on response of single units of the preoptic
anterior hypothalamic areas, local, midbrain and spinal cord temperatures: a, network for hyperthermic activity (warm); b, network
for hypothermic activity (cold). (Hardy and Guieu, 1971).

stimuli as demanded by their models. These neurons, which respond only to the
local temperature changes, are shown in Figures 14a and 14b as furnishing independent temperature inputs into the networks. Interneurons of the logical scheme
respond to many stimuli including temperature changes in the midbrain reticular
formation (T_{mb}), spinal cord (T_{sc}), skin (T_s) and viscera (T_{re}). The fitting
process was greatly simplified by assuming two separate but interacting networks,
one effective in hyperthermia (warm) and the other in hypothermia (cold). The
outputs, W_1, W_2, W_3, W_4 and C_1, C_2 can be combined to serve the physiological
functions of temperature regulation as shown in figure 15; the outputs noted as
U_w and U_c were omitted from the scheme because they responded to both heating
and cooling in the same way. The similarity of figure 15 to the models of Bligh,
Hammel and others is clear with the exception of cutaneous vasomotor control.

At the present time even a restricted neuronal model such as shown in figures
15 and 16 must be viewed with caution, but the possibilities for challenge are
attractive.

The final form of model I wish to mention is the chemical or neural transmitter model.

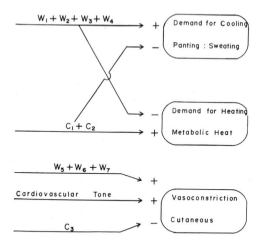

Fig. 15 - Model representing hypothetical summations of outputs of cold and
warm neuronal networks.

Chemical Models

In 1964, Feldberg and Myers advanced the idea that body temperature is basically controlled by a balance of the neural transmitters 5HT and nor-epinephrine.
Their concept was based on the fact that these substances are found more concentrated in the preoptic-anterior hypothalamus than elsewhere in the brain, and that
ventricular perfusions containing these substances could induce hypo or hyperthermia. Recently (Myers, 1970) a second chemical factor has been proposed based on
experiments in which the Na-Ca ratio is changed in the posterior (but not the anterior) hypothalamus. The latest representation of this system, shown in figure
16, was presented to the CIBA symposium on pyrogens. In a sense this model is a
neuronal circuit similar to that proposed by Hammel although the ideas of Hans
Meyer's model of "two centers" are also included. Peripheral thermal stimulation
is shown leading directly to the anterior hypothalamus; here, via transmitter
substances 5HT and NE, they synapse with temperature sensitive neurons and other
units of the "aminergic thermostat". The preoptic-anterior hypothalamic area is
visualized as the place of action of pyrogens, drugs, temperature change, etc.
and the resulting activity is transmitted via a third transmitter substance, ACh,
to the posterior hypothalamus. In the posterior hypothalamus, the sensitivity
to pyrogens, temperature, etc. is lost but the "setpoint" temperature of the
body is added through the control of the Na-Ca ratio.

This model is pictorial and descriptive but its simpler predecessor from 1964
has stimulated much research in various laboratories in the last few years. At
present, there is considerable confusion in this area since agreement on basic
data has not been achieved, and the conflicts in results are obscuring our vision.
However, the importance of the area is unquestioned and the possibilities of challenge to the Feldberg-Myers model are indeed interesting.

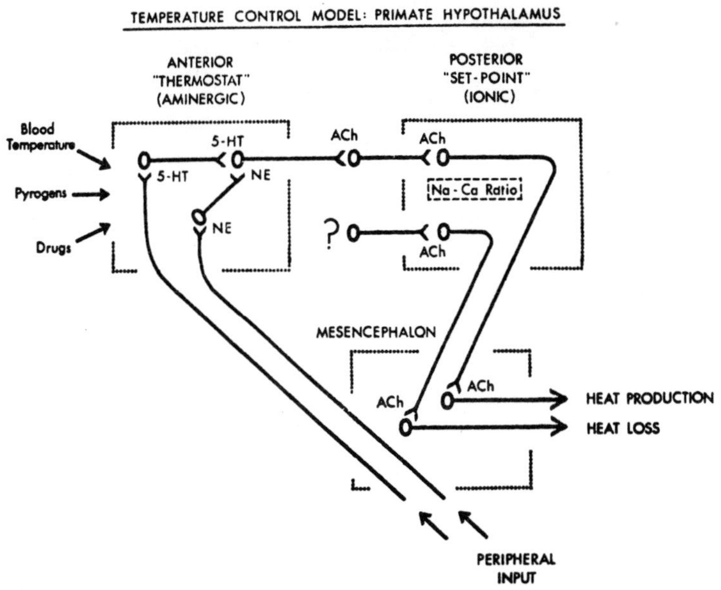

Fig. 16 - Chemical model indicating peripheral control and posterior hypothalamic
        "ionic setpoint" (Myers, 1970).

A chemical model which has much physiological data supporting it is that proposed by Bligh in the essay in this volume. Using a combination of ambient
temperatures and centrally administered neural transmitters, Bligh and his colleagues have clarified many of the earlier findings in a number of species and
have offered a model of the efferent network which has wide acceptance. In
some form, Bligh's concepts have been incorporated into most formulations of
the thermoregulatory system, as seen in Figure 15 above and essays by Hammel,
Mitchell and Cooper.

Other chemical models exist at this time to deal with particular subsystems of
the biological thermoregulator. For example, Carlson and Hsieh (1970) have a
model to account for the peripheral vasomotor control exercised by the sympathetic nervous system, and Brück and Wünnenberg (1970) proposed a model for
simulating the brown fat organ to produce metabolic heat in the cold. These
are examples of the detail in which thermoregulatory activity is being studied
and how far the field has moved in the last 20 years. So, what is the conclusion about models of thermoregulation? Have we made use of those we have?
To answer this question, I present a slight modification of our scientific mill
in figure 17, by including a box indicated as "experimental challenge" in the
"intuitive test loop". That is, although comparison of our analytical hypotheses
(however they may be expressed) with data from the real world may move us to alter
our concepts, it seems clear that physiologists use models as points of departure
in formulating intuitive challenges. Thus, new experimental data become avail-

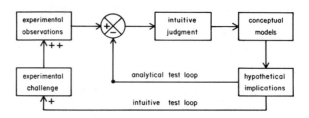

Fig. 17 - Modified servosystem for research.

able requiring models to be changed. As long as this process continues, the models, in a sense, feed on themselves in a positive feedback loop. Analytical models are therefore not necessary but they can be useful and thermophysiologists have used them effectively for a long time.

Do we have too many models? The answer to this question must be no - we don't have nearly enough to represent the many subsystems involved to say nothing of the overall system performance. It seems likely that we will have increasing dependence on models of some sort in the future but that greater care will be needed in their preparation and deductions from them will have to be tested out to some degree before announcement. A model which falls to the first challenge should probably not have been formally proposed; thus, models may promote rigor in our thinking as well as in our laboratory procedures.

REFERENCES

Adair, E.R. (1972) Evaluation of some controller inputs to behavioral temper-
    ature regulation. Int. J. Biometeor. (In press)
Aschoff, J. and R. Wever (1958) Modellversuche zum gegenstromé - Warmeaustausch
    in der Extremitat. Zeit. ges. exp. Med. 130, 385.
Bazett, H.C. (1927) Physiological responses to heat. Physiol. Rev. 7, 583.
Bazett, H.C. (1949) The regulation of body temperatures, Chapt. 4, In: Physiology
    of Heat Regulation and the Science of Clothing. W. B. Saunders Co. Phil.
Benzinger, T.H. (1959) On physical heat regulation and the sense of temperature
    in man. Proc. Nat'l. Acad. Sci., 45, 645.
Benzinger, T.H. (1963) Peripheral cold and central warm-reception, main origins
    of human thermal discomfort. Proc. Nat'l. Acad. Sci., 49, 832.
Bernard, C. (1865) Introduction à l'étude de la médicine expérimentale.
    Bailliere, Paris.
Brück, K. and W. Wünnenberg. (1970) "Meshed" control of two effector systems:
    nonshivering and shivering thermogenesis. Chapt. 38, p. 562 in Physiological
    and Behavioral Temperature Regulation, Chas. C. Thomas, Publ., Springfield,
    Ill.
Burton, A.C. (1934) The application of the theory of heat flow to the study of
    energy metabolism. J. Nutr., 7, 497

Carlson, L.D. and A.C.L. Hsieh (1970) Control of Energy Exchange. The MacMillan
   Co., New York, NY
Chatonnet, J. and M. Cabanac (1965) The perception of thermal comfort. Int. J.
   Biometeor. 9, 183.
Crosbie, R.J., J. D. Hardy and E. Fessenden (1961) Electrical analog simulation
   of temperature regulation in man. Inst. Elec. Electronic Engineers, Trans.
   of Biol.-Med. Electronics, 8, 245.
DuBois, E.F. (1948) Fever and the Regulation of Body Temperature. Chas. C.
   Thomas, Publ., Springfield, Ill.
Fanger, P.O. (1967) Calculation of thermal comfort: Introduction of a basic
   equation. ASHRAE Trans., 73, p. 1.
Feldberg, W. and R. D. Myers (1964) Effects on temperature of amines injected
   into the cerebral ventricles. A new concept of temperature regulation. J.
   Physiol. (London) 173, 226.
Gagge, A.P., J.A.J. Stolwijk and Y. Nishi. (1971) An effective temperature scale
   based on a simple model of human physiological regulatory response. ASHRAE
   Trans., 74, 1.
Grodins, F. (1970) Theories and models in regulatory biology, Chapt. 49 in
   Physiological and Behavioral Temperature Regulation, Chas. C. Thomas, Publ.,
   Springfield, Ill.
Hammel, H.T. (1965) Neurons and temperature regulation, Chapt. 5, p. 71 in:
   Physiological Controls and Regulation, W. B. Saunders Co., Phil.
Hammel, H.T., J. D. Hardy and M. Fusco (1960) Thermoregulatory responses to hypo-
   thalamic cooling in unanesthetized dogs. Am. J. Physiol., 198, 481
Hardy, J.D. (1971a) Thermal comfort and health, Am. Soc. Heating, Refrig. Air
   Cond. Engineers, ASHRAE Journal, 77, 43.
Hardy, J. D. and E. F. DuBois (1938) Basal metabolism, radiation, convection and
   vaporization at temperatures of 22 to 35°C. J. Nutr. 15, 477.
Hardy, J.D. and J. D. Guieu (1971b) Integrative activity of preoptic units, II.
   J. Physiol. (Paris), 63, 264.
Hellon, R.F. (1971) Hypothalamic neurons responding to changes in hypothalamic
   and ambient temperatures. Chapt. 32, p. 463, in Physiological and Behavioral
   Temperature Regulation, Chas. C. Thomas, Publ., Springfield, Ill.
Hensel, H. (1952) Physiologie der thermoreception. Ergeb. Physiol. 47, 166.
Hill, L. (1923) Kata-thermometer in studies of body heat and efficiency. Med.
   Res. Council of Great Britain, Special Report Series 73, 48.
Liebermeister, C. (1871) Untersuchung über die quantitätizen deraveränderungen
   Kohlensaure Production dem Menschen. Deut. Arch. Klin. Med., 8, 153.
MacDonald, D.K.C. and C. H. Wyndham (1950) Heat transfer in man. J. Appl. Phys.,
   3, 342-364.
Meyer, H. H. (1913) Theorie des Fiebers und seiner Behandlung. Verhandl. deut.
   Bes. inner Med. 30, 15.
Myers, R.D. (1970) Hypothalamic mechanisms of pyrogen action in the cat and
   monkey. Ciba Symposium on "Pyrogens and Fever", Churchill Ltd., London.
Nakayama, T., J.D. Eisenman and J. D. Hardy (1961) Single unit activity of
   anterior hypothalamus during local heating, Science, 134, 360-361.
Ott, I. (1887) Heat center in the brain. J. Nervous and Mental Disease, 14, 152.
Ott, I. (1904) A Textbook of Physiology, p. 348, F.A. Davis Co., London.
Pickering, G. (1958) Regulation of body temperature in health and disease.
   Lancet, 4, 1.
Richet, C. (1885) Die Beziehung des Gehirns zur Körper wärme und zum Fieber.
   Arch. ges. Physiol., 37, 624.
Rubner, Max. (1902) Die Gesetze des Energieverbrauchs bie der ernahrung
   Leipzig and Vienna, Franz Deutiche.
Smith, P.E. and E. W. James (1964) Human responses to heat stress. Arch of
   Environ. Health, 9, 332.
Stolwijk, J.A.J. (1971) A mathematical model of physiological temperature regu-
   lation in man. NASA CR-1855, U.S. Nat. Tech. Infor. Service, Springfield,
   Va. 22151.

Stolwijk, J.A.J. and J. D. Hardy (1966) Temperature regulation in man, a theoretical study. Pflügers Archiv. 291, 129.

Thauer, R. (1967) Contribution à l'étude de récepteurs profonds, extra-crânieus, du froid chez les homéothermes. Arch. Sci. Physiol. 21, 205.

Vernon, H.M. (1932) Proc. Inst. Heating and Ventilating Engineers (London) 31, 100.

Waites, G.M.H. (1961) Polypnea evoked by heating the scrotum of the ram. Nature (London) 190, 172.

Wissler, E.H. (1961) Steady state temperature distribution in man. J. Appl. Physiol., 16, 734.

Wyndham, C.H. and A.R. Atkins, (1968) A physiological scheme and mathematical model of temperature regulation in man, Pflügers Arch., 303, 14-30.

SUBJECT INDEX